Parasitic Weeds of Jordan: Species, Hosts, Distribution and Management

(Part 1: Root Parasites; Orobanchaceae, Santalaceae & Cynomoriaceae)

Authored by

Jamal Ragheb Said Qasem

(Professor of Weed Science)
Department of Plant Protection,
Faculty of Agriculture
University of Jordan,
Amman, Jordan

Parasitic Weeds of Jordan: Species, Hosts, Distribution and Management

(Part 1: Root Parasites; Orobanchaceae, Santalaceae & Cynomoriaceae)

Authored by

Jamal Ragheb Said Qasem

Professor of Weed Science
Department of Plant Protection,
Faculty of Agriculture,
University of Jordan,
Amman,
Jordan

Parasitic Weeds of Jordan: Species, Hosts, Distribution and Management

Part 1

Root Parasites; Orobanchaceae, Santalaceae & Cynomoriaceae

Author: Jamal Ragheb Said Qasem

ISBN (Online): 978-1-68108-874-7

ISBN (Print): 978-1-68108-875-4

ISBN (Paperback): 978-1-68108-876-1

need for a court order if at any point you breach any terms of this License Agreement. In no event will any delay or failure by Bentham Science Publishers in enforcing your compliance with this License Agreement constitute a waiver of any of its rights.

3. You acknowledge that you have read this License Agreement, and agree to be bound by its terms and conditions. To the extent that any other terms and conditions presented on any website of Bentham Science Publishers conflict with, or are inconsistent with, the terms and conditions set out in this License Agreement, you acknowledge that the terms and conditions set out in this License Agreement shall prevail.

Bentham Science Publishers Ltd.
Executive Suite Y - 2
PO Box 7917, Saif Zone
Sharjah, U.A.E.
Email: subscriptions@benthamscience.net

BENTHAM SCIENCE

CONTENTS

PREFACE

I preferred to start this book with a statement from one of our local farmers who owns an olive farm heavily attacked by the mistletoe (*Viscum cruciatum*) in the northern part of Jordan (Mediterranean biogeographical region, although our knowledge regarded the parasite as a tropical and subtropical). In the farmer's reply on whether he knows what appears heavily covering/attacking his trees? He simply replied, "This is the cancer of the trees". In his brief answer, the farmer summarized the whole story of parasitic weeds. To this end, we should deeply think in the farmers' words, why and how? Certainly, he intended to say that there is nothing I can do to relieve or save these trees, but I am only waiting for their death and removal from the field. The farmer actually summarized the long lasting research work history since one and a half-century on these parasites and all aspects. This conveyed that until now, the large volume of literature available on these species just brought a little but not an effective and absolute satisfying solution, while losses in plant growth and yield are extensive and brought total death and loss in most cases. It is worth mentioning that broomrapes (*Orobanche* spp.) are called "Halook" in Arabic, which means exhaustive and destructive.

Parasitism is an unusual behavior difficult to understand or justify the reason/s behind it without rigorous research work on all aspects of this phenomenon. Parasitic habit is against natures' law that offers opportunities for individuals, populations, communities to obtain some or all of their needs to grow and exist. Therefore, competition may be the only logical mechanism in struggling for resources in nature at which the weak competitor is limited in growth or space it occupies or extinct. Parasitism is a different mechanism of negative interaction between plants in nature at which one organism depends and lives at the expense of others for food and/or water for part (hemi-parasites) or the entire life span (holo-parasites). Therefore, regardless of the quantity or quality of the resource/s captured by the host plant but these must be shared, mostly or totally with the parasite on the expenses of host species captured these from the environment. The prey has no choice but to accommodate itself accordingly and to cope with parasite aggression, although suffering partial or total death. The power of parasitism and the philosophy behind this behavior is not fully understood since exerted by an individual or a group of individuals on others of their kind (plant on another).

What is actually known till now is just a little on the behaviour and story of these parasites and thus any chemical or control/management technique must consider the interrelationship between the parasite and its host and possible injury to host plants, their normal growth, performance, physiology, biochemistry, existence, the value of the treatment and the final yield of treated plants compared with parasite-infected but untreated plants in both growth and yield obtained. All these bring to mind the Ayat in The Holy Quran that says, "They ask you about "the spirit."Say: "The spirit descends by the command of my Lord, but you have been given only a little knowledge."Quran, Surah Al-Isra ayat 85. Parasitic flowering plants or parasitic weeds are most destructive to many cultivated crops and wild plant species of all other weed species we observe in our fields, although not attacking all plant species but their hosts. However, parasitic weeds attack plant species differing in growth habit and habitat and range between forestry climax species down to sand fixer desert shrubs and wild herbs as well as cultivated crops under different conditions. These are unique in relation with their hosts and peculiar in behaviour in all aspects at which every single issue in physiology, biology, ecology and behaviour is only a small part of a more general adopted strategy that needs to be deeply investigated and studied in order to understand the real story that stands behind it. Research on parasitic species requires long term experience, high training and following upon recent developments and literature on worldwide achievements on the subject and parasites persistence. The volume of literature available on parasitic weeds is a huge and tackles all aspects in relation to these species with the hope of finding some information that can lead to better understanding, partial solution or some knowledge on these species toward managing their spread and destructive effects on agricultural and ecosystems at a wider scale. To update, all factors appeared to play or have a positive impact on these species and their existence while increases in the technological advancement in control, including recently developed herbicides or control techniques are still far from reality or partially solving the problems these cause to farmers and exert on the agro-ecosystem worldwide. However, management of parasitic species would never be possible without understanding their biology, physiology, biochemistry, ecology, distribution and genetics, as well as their external and internal nature of relations with host plants. All must be well considered in any integrated control program. The total number of parasitic weeds is approaching 5000 species worldwide, very well equipped with certain characters that enable their spread and existence, although many are not in direct contact with the soil factors needed for germination and growth. Some of these characters include the huge number of seeds produced and long dormancy, host germination stimulant and preconditioning period required for germination of certain parasites, seed dispersal by non-specialised agents as well as through an active dispersal mechanism such as for mistletoes and *Osyris*. Adaptation to long and short dispersal distances that lead to the high vertical or horizontal spread of some of these parasites, ability to propagate sexually by seeds or/and vegetatively through the suckers embedded in host tissues for certain parasites (such as dodders, mistletoes and osyris) and their high compatibility with host plant physiology and biochemistry. However, the great number of these parasites in the world and their prolific seed production, along with partial or total dependence on host plants for food and nutrients along with their active dispersal by specialized and nonspecialized agents, these associated with the absence of effective control measures including high technological measures all contributed in the wide spread magnitude and the destructive effects these cause to a wide range of host species attacked ranged from desert forage and range land species found on poor sandy soil under severe drought conditions (*Cistanche* spp.) to species found growing under a high average annual rainfall and on rich clay organic soils as well as under rainfed and irrigated conditions (*Orobanche* spp.) and parasitic species in tropical and subtropical regions as well as in the Mediterranean's such as mistletoes. However, more extensive, rigorous research work is still needed to counteract partially or significantly the destructive effects these parasites cause to agriculture and the environment. Intensive work should be directed toward better understanding the nature of host-parasite relationships, which must go with extensive research work devoted to their management through more general integrated weed control programs based on recent research developments. Therefore basic and applied research work must go in parallel while integrated parasite control programme should always be considered in managing these species and reducing or eliminating their negative impact on the agro- and eco-systems.

In the present book, 8 parasitic genera of 6 plant families were included and discussed. Some of these are among the important worldwide parasites, such as broomrapes, dodders and mistletoes. I hardly tried to include basic information on parasitic families that occur in Jordan in order to help readers gain knowledge on these parasites and their influences on vegetation, environment, food security, and human welfare. Species included under each family were dealt with separately and basic information on each was provided along with updated research results available on their control or management in general. The book illustrates coloured photos (where possible) of each parasite with its hosts while tables included parasitic species and their hosts with necessary information were presented. As an author with almost half a century of research work devoted to weed control, including these parasites and mainly *Orobanche,* I feel that knowledge gained throughout the whole period is fair, while great information is still unknown on simple issues in relation to these parasites. We know that competition between individuals of certain parasitic species such as *Orobanche* on the root system of its host is evident and may be responsible partly for the differences in the development of parasite individuals on the host roots that range from a small haustorium to emerged well developed parasite shoot and similar variations that could be easily observed in mistletoe (*Viscum* individuals) subsisted on a single shoot of the host plant at which all stages can be observed at once on the same shoot or stem and range from germinated seed up to a fully developed and even fruiting parasite. However, this parasite exhibits self-parasitism, which is unusual and unpredictable. Therefore, whatever the achievements reached in research on these species and their deep strategy adopted for existence and continuity of their genetic lines is still not fully understood and thus we can't claim that their control is possible in the near future with the existing information and available literature although extensive. Much in-depth and well-designed work is still needed and for all to learn and understand our local farmers' reply on the mistletoes attacking his olive trees which has summarized the whole story till now. I hope and ask God that literature and illustrations presented in this book would provide readers, researchers, farmers and students with some useful information on parasitic species included and the problems they cause to agro-ecosystems although can be of some positive values of medicinal, ecological, ornamental and scenery to human society but for sure regardless of their general negative impacts, they form a significant component of the agroecosystem that could influence vegetation, soil cover and food production.

Jamal Ragheb Said Qasem
Professor of Weed Science
Department of Plant Protection
Faculty of Agriculture
University of Jordan
Amman, Jordan

ACKNOWLEDGEMENTS

All thanks to God, who encouraged, directed, and helped complete this book to find it a reality in the world libraries. I want to express my deep thanks to the Deanship of Scientific Research at the University of Jordan, who funded two research projects on parasitic weeds and to the Higher Council for Science and Technology who approved and funded a project on the same subject through which it was possible recording and photographing the parasitic weeds in the Jordan flora and record their spread and hosts in the country. I would like to thank the research assistants who worked on these projects. Thanks to Bentham publishing family for printing and publishing the book. To my wife and sons who supported and encouraged me throughout the whole period, all my love. Finally, to all colleagues and friends worldover who their hard work was always in mind, remembered and encouraging when to read their publications and was encouraging to keep on the same level, I say thanks, your active research work and publications were behind my persistence and insistence to keep on the same level and have this book completed. Having realized the merits of its publication, I offer it to all who are homesick and in love with scientific research, who devote life to science, homelands and believe in history and originality. Although a large volume of congress proceedings, literature, and reference books are available on parasitic weeds updating seems necessary to stay familiar with recent developments on the subject. I hope this book is another brick in the long science wall for the scientific community, students, researchers, and those interested in parasitic weeds.

CONSENT FOR PUBLICATION

Not applicable

CONFLICT OF INTEREST

The author confirms that there is no conflict of interest.

DEDICATION

- In the name of God, who created these parasites and their hosts and challenged knowledge and Science of all. God, you didn't create these and others aimlessly.
- To my homeland where these parasites and their hosts prevail and are offered to all to learn the secrecy of their life, existence and natures' law.
- To the memory of my parents, who made sacrifices for my education and taught me that the land is faith and life.
- To my students and those who seek knowledge, to search, learn and understand some complexicity in plant-plant negative interactions.

ABSTRACT

Parasitic weeds or parasitic flowering plants are destructive species that negatively impact agro- and ecosystems since parasitize cultivated and wild grown annual, biennial and perennial plant species of different growth habits and habitats. Parasitic weeds attack most vegetables, forage plants, shrubs, and fruit and forest trees, reflecting a high physiological/chemical and environmental tolerance and/or these parasites adapt to separate plant cells, dissolve tissues, enhance haustorium penetration and thus secure food and water availability for their own use. Parasitic weeds are unusual and peculiar species in growth habits and habitats, hence not in direct contact (except certain hemi-parasites) with soil and soil factors necessary for growth, development and existence and for almost the entire life cycle. Instead, they are usually attaching themselves to other plant species of different genera (except self-parasitism) and families through an absorptive organ called haustorium (pl. haustoria) that develops upon attachment to host tissues. Essentially all haustoria are modified roots (Nickrent, 2002), sucker like structures varied in size, shape and morphology for different parasitic species and through which the parasite can derive its food and /or water from host plant. Parasitic weeds are of widespread significance in different parts of the world and a great concern to farmers and researchers. Certain species are a serious threat to food security since parasitize many food crops of economic importance. However, parasitic weeds are different in requirements for germination, attachment to host plants and host parts attacked, degree of dependence upon hosts, and presence or absence of chlorophyll. Most recent management methods are presented including (i) biological using natural enemies such as fungi, bacteria and insects including *Trichoderma, Phytomyza* for *Orobanche*; *Smicronyx* sp and fungal species for *Cuscuta* and *Orobanche;* (ii) Chemical by the use of herbicides such as glyphosate for *Orobanche* control in fababean and different sulfonylurea herbicides for other species (e.g. *O. aegyptiaca* and *O. ramosa* in tomato) and selective application of MCPA and 2,4-D for mistletoes control; (iii) physical including soil solarization and plastic mulch for *Orobanche* and *Cuscuta* species; (iv) organic methods using plant residues and natural products or allelopathy (e.g. for *Orobanche*), root extracts (sunflower), plant oils of ginger, groundnuts, palm, sunflower, safflower, castorbean, linseed, neem, coconut or tobacco seed oils, niger (*Guizotia abyssiniaca*), and mustard oils, orobanchol and alectrol germination stimulants, different strains of *Streptomyces,* certain fungal metabolites including cotylenins and fusicoccins; (v) trap and catch species in crop rotation or intercropping systems ; (vi) genetically engineered crops (herbicide resistant); (vii) plastic mulch with fertilizers (chicken manure, urea, ammonium sulphate); (viii) resistant cultivars for different crop species and screening studies on tolerant/resistant crop lines; (ix) mowing and general contact herbicides such as paraquat and dinoseb for *Osyris* and *Cuscuta* (early infestation on perennial field crops) species control; (x) pruning and shading for mistletoes control; and (xi) integrated control methods. This book reviews parasitic weed genera that occur in Jordan and are of great importance to many other Middle East countries. It includes recent research achievements in combating these parasitic weeds and presents a brief description of species phonology, ecology, and economic importance as parasites of cultivated land, forestry and wild species in the country. Recent literature on species biology, physiology, and ecology and on developments on their control and management were reviewed for each parasitic genus separately as deemed necessary.

Key words: Parasitic weeds, Biology, Physiology, Ecology, Control methods.

GENERAL INTRODUCTION

Parasitism is a phenomenon of widespread importance among angiosperms and there are at least five orders having parasitic species (Musselman, 1982) that evolved independently in 20 plant families (Musselman, 1987) and comprising 3000-5000 species (Musselman, 1987; Sauerborn, 2001). Almost 4200 species of haustorial parasitic plants are spread in the world, representing just over 1% of all flowering plants. Nickrent (2002) reported that the total number of parasite species is not distributed evenly among the 18 families or 274 genera. Aly (2007) reported that about 20 families (3,000–5,000 species) of higher plants are parasitic in the plant kingdom and may cause production losses of 30–80% in staple food and industrial crops in every continent. Westwood *et al.* (2010) mentioned parasitic weeds include approximately 3,900 species included in more than 20 plant families consist the important genera *Striga, Orobanche* and *Cuscuta.* Fernández-Aparicio *et al.* (2011) reported at least 4500 plant species within 270 genera in over 22 families predominantly angiosperms rely on a parasitic association with a host plant for their mineral nutrition, water and/or carbon supply. Heide-Jorgensen (2013) mentioned that the total number of world parasitic plants account for 4500 species present in 275 genera of 28 botanical families. However, in a recent molecular study, Nickrent (2020) reported the occurrence of 4700 species that belong to 290 genera of parasitic angiosperms. Approximately 30 genera of these parasites have been reported to negatively impact cultivated crops, while only about 11% of all genera have members that could be considered pathogenic. Parasitic species belong to different plant families and are either root or shoot parasites, attack plants of different botanical families and growth forms and are distributed in different geographical regions of the world. They cause great yield losses and a heavy infestation can negate yield and lead to complete crop loss. Some parasites are regarded as highly damaging agricultural pests that exert significant threats to regional and global commodity and food production (Cudney *et al.*1992; Aly 2007; Parker 2012). However, out of plant families that have parasitic species, six were considered of great economic importance, including Scrophulariaceae, Orobanchaceae, Santalaceae, Cuscutaceae, Viscaceae and Loranthaceae. Other researchers also added Lauraceae and Balanophoraceae (Subhashini *et al.*, 2019). In general, *Cuscuta*, *Arceuthobium*, *Orobanche*, and *Striga* were regarded as causing the most damage to economic host crops.

Parasitic weeds are quite varied in their dependence upon host plants. Some are holo-parasites (complete parasites) devoid of chlorophyll (or nearly so); and completely dependent on hosts for organic carbon, nutrients and water, and most are root parasites. Certain stem parasites (e.g. *Cuscuta europaea*) have lost rubisco, thylakoids, chlorophyll and light-dependent CO_2 fixation (Machado and Zetsche, 1990), while others have green pigments, do photosynthesis (at least during some portion of the life cycle) and can produce their own food (hemi- or incomplete parasites), but also parasitize other plant species from which they partially or totally obtain their food and/ or water. These can be further divided into facultative (do not require a host to complete life cycle) such as *Osyris* and *Thesium*, and obligate completely host- dependent to flower and seeding or fruiting such as *Viscum* and *Loranthus* spp. However, all parasitic species appeared to lack a normal root system for almost all growth stages after germination, and instead they have absorptive organs or structures that give physical support to the germinated seedlings attached to the host tissues. Certain parasitic species subsist themselves on the host root systems while others are stem or shoot parasites and thus represent a highly ecological, biological and physiologically tolerant species to both internal and external environmental changes.

Parasitic species are different in seed production, longevity, disposability and dispersing agents. Seeds are well adapted to disseminate by different general (unspecialized) agents to survive and tolerate severe environmental condition under which long-term ecophysiological dormancy has been very well demonstrated and developed in many parasitic species in the absence of suitable conditions for germination, mainly the availability of host plants. Birds and animals are the main dispersing agents for *Loranthus*, *Osyris* and *Viscum,* but wind, contaminated crop seeds and packed agricultural materials are responsible for *Cuscuta*, *Cistanche*, *Orobanche* and *Thesium* seed dissemination.

Variations in host range within and among parasites are evident (Parker and Riches, 1993), depending on the chosen co-evolutionary pathway of these parasites (Musselman, 1987), while host preference for the same species or between species of the same or different genera has also been documented (Parker and Riches, 1993; Qasem and Kasrawi, 1995; Hudu and Gworgwor, 1998). However, differences between parasitic weeds in their requirements for host stimulants resulted in a wide variation in the range of their hosts. Certain parasitic weeds are mainly confined to specific hosts and appear highly specialized to woody shrubs, fruit trees or forestry species (*Loranthus, Viscum*, *Osyris*, *Cistanche*); others are more restricted to herbaceous and field crops (*Thesium*, *Striga*), while a third group attacks both woody and herbaceous species (certain species of *Orobanche* and *Cuscuta*). Differences between species of the same genera in their host range have also been noted. Certain parasitic species cannot germinate unless the pre-conditioning period and host stimulants are fulfilled (*Orobanche*, *Striga*, *Cistanche*), but others (*Cuscuta*, *Thesium*, *Osyris*, *Viscum* and *Loranthus*) are not in need of such conditions but appeared in a limited and specific host range, and not attacking other plant species, reflecting host preference/specificity or certain host mechanical, physiological or chemical resistance. The conditioning period may enhance the responses of dormant seeds to germination stimulants of host plants and shorten the response period. However, *Orobanche* species may be different in their need for such a conditioning period at which certain species do not require preconditioning for germination or response to germination stimulants (Plakhine *et al.*, 2009). Non-conditioned seeds of both *Orobanche cumana* Wallr. and *O. aegyptiaca* Pers. (syn. *Phelipanche aegyptiaca* (Pers.) Pomel) were able to germinate in response to chemical stimulation by GR24 even without prior conditioning. The mechanism behind the failure of these parasites from attacking other plant species is unknown and may need investigation but could be utilized as another method of their control. Different workers mentioned that host preference in *Striga* spp. is due to different germination stimulants in the host root exudates (Bharathalakshmi and Jayachandra, 1984; Bharathalakshmi *et al.*, 1990). Host-parasite interaction at germination time is very specific and depends on chemical recognition. Several natural germination stimulants have been isolated from hosts (Galindo *et al.*, 2004). With one exception (sorgoleone), all germination stimulants isolated so far from host plants belong to strigolactones. Strigolactones have been isolated and identified as germination stimulants from sorghum (Hauck *et al.*, 1992; Siame *et al.*, 1993), red clover (Yokota *et al.*, 1998) and cowpea (Muller *et al.*, 1992).

The ability of host and nonhost plant species to stimulate parasite seed germination or inhibit attachment encouraged studies on the possible use of these as catch or trap species to reduce parasite infestation. Stimulants of root parasites are mainly natural chemicals (Allelochemicals) released through the host's root system into the surrounding environment and received by parasite seeds. These may stimulate parasite seed germination by affecting internal hormonal balance and overcoming the special seed chemical-based dormancy. They may also act as hormones (e.g. strigols and their analogs of GA-like chemicals) affecting the embryo, germination and growth of the parasite germination tube. Other chemicals produced by host plants could stimulate stem parasites to penetrate and grow inside host tissues (Joel and Goldman-Guez, 2001). It is worth mentioning that both susceptible and tolerant or resistant species produce germination stimulants, but differences exist in the amount of stimulants they produce and received from parasite seeds. However, some of the chemicals exuded from roots of resistant, tolerant or non-host species may contain phytotoxic materials that prevent seed germination, harm germinated parasite seeds, or prevent seedling attachment. In this regard, it has been reported that xenognostic allelochemicals that trigger haustoria development are commonly quinones, phenolic acids, or flavonoids derived from the phenylpropanoid pathway (Riopel and Timko, 1991; Albrecht *et al.*, 1999). The relationship between parasites and their hosts is highly complicated, both may work together for a successful attachment and complete parasitism, but the antagonism is most common between both at which different mechanisms may be involved in the resistance of host to parasite germination, attachment, penetration and development which is termed as tolerant/resistance mechanism or sometimes described as a defense mechanism. The parasite may be well prepared for apositive response to host signals and show the ability to attach itself to the host cell surface and to penetrate with theaid of enzymes or other chemical secretions,which allows dissolvent and separation of host cells. In the centre of these complicated chemical reactions between parasites and host plants, allelochemicals are involved. Both partners may use these chemicals to negate the influence of each other, either by stimulation of parasite and host growth, or each may use these as harmful natural chemicals to achieve their ultimate goal and thus in some cases, facilitate association (Qasem, 2006b). However, the stage from seed germination to attachment is the most sensitive in the life cycle of parasitic weeds and control measures should be applied during this phase (Salle *et al.*, 1984; Dhanapal, 1996b).

Differences in the attachment of different parasitic species and their ability to resume growth after being mechanically removed from host plants have also been noted. Certain species such as *Orobanche*, *Cistanche*, *Cynomorium* and *Thesium* are weakly connected to host tissues and cannot resume growth after being detached from host plants; others such as *Cuscuta*, *Viscum*, and *Loranthus* can re-vegetate and resume growth from the suckers or haustoria embedded in host tissues. This may reflect differences in reproduction methods and the life cycles of different parasitic species. However, all are referred to as annuals, reproduce by seeds, although certain species are able to vegetatively reproduce from the overwintering haustoria inserted in host tissues (certain *Cuscuta* species and *Loranthus*) and behave as perennials.

Differences between parasitic species in seed production and germination, attachment to host plants, morphology and growth habit, physiology and anatomical relationships between parasites and their hosts have created serious control difficulties. Selective chemical control using herbicides must be well managed, although it is far from a success and difficult to achieve in most cases. Herbicide selectivity mainly depends on differences in tolerance between the parasite and host plants, while effective rates of herbicides may injure host plants. In addition, certain parasitic species are only observed after emergence on the soil surface (e.g. species of *Orobanche, Cistanche, Cynomorium* and *Striga*), while most damage to host plant occurs below the soil and before parasite emergence. However, different control methods are recommended and employed against many of these parasites, while some other promising results have also been reported.

Once the parasite becomes in contact with the host plant, it starts forming and developing a sucker like structure (haustorium) inserted into host tissues (Fig. 1). Haustoria are varied in size, length and number per individual parasitic plant depending on the species and range in number between one (single connection point) such as for *Orobanche*, *Striga* and *Viscum* to many in *Cuscuta* and *Loranthus* (Fig. 2).The haustorium is a bridge of tissue connecting host and parasite; it is usually a swollen mass consisting of both host and parasite tissues and act as a channel for the flow of water and nutrients from host to parasite. The chemical relations between parasitic species and their hosts can be described as a compatible/antagonistic one since both involve chemicals that stimulate germination and attachment, and other incompatible, phytotoxic that prevent germination, attachment, growth or development of the parasite. The first group can be classified as synergistic chemicals or stimulants, while the second is termed antagonistic or inhibitory.

Mallet (1973) suggested that root extract of the broad bean contains both inhibitory and stimulatory substances for *Orobanche crenata* Forskal. seed germination and Whitney (1978) came to the same conclusion and clearly showed that certain nonhost species to *O. crenata* and *O. ramosa*

Orobanche ramosa *Viscum cruciatum*

Viscum cruciatum *Osyris alba*

Fig. 1. Different parasitic species showing haustoria developmental stages.

release compounds that inhibit haustorium at different development stages of parasite germination even in the presence of suitable stimulants and the possibility that different stimulants may interact together in inducing seed germination of the same parasite also exists. Both types of chemicals, however, were considered as allelochemicals (Rice, 1984) and later based on the International Allelopathy Society (IAS) definition of Allelopathy and therefore both stimulation and inhibition processes are involved in plant-plant interactions and the whole phenomenon was classified and included under the term allelopathy (IAS Constitution and Bylaws, Gadiz, Spain, 1996). The wide differences between species in the host range proved differences in their ability to produce different tissue-dissolving enzymes or their fitness and compatibility with indigenous chemicals that host plants release in the form of phytoalexins as a defense mechanism against parasite attack and invasion. Differences between host plants in the production of defense chemicals or stimulating parasite attachment explain in part the susceptibility or tolerance level

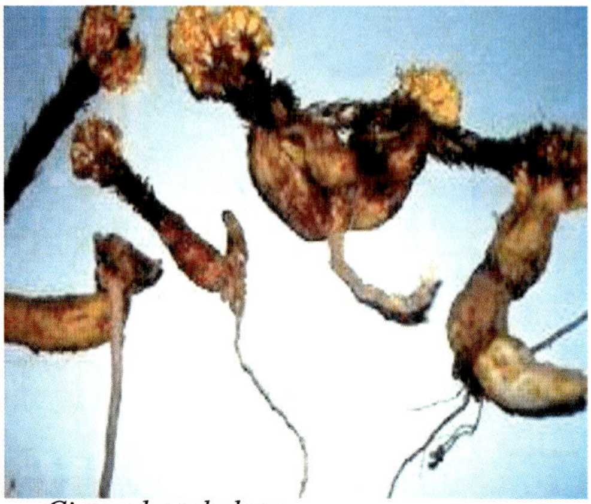

Loranthus acaciea *Cistanche tubulosa*

Fig. 2.*Loranthus acaciea* (many connection points) and *Cistanche tubulosa* (a single connection point)

that these species or cultivars show in response to parasite invasion. Joel and Goldman-Guez (2001) reported pectin in host cell walls de-esterify, presumably by enzymes derived from the host itself, indicating that some steps in penetration are facilitated by the host through a mechanism induced by the parasite. It has also been suggested that *Striga* lactase activity releases quinones from the host root that stimulate haustoria initiation (Stewart and Press, 1990). In this regard, different studies tried to quantify the relationship between host species and their parasites and the reflection on their general ecological responses. In a study to analyze the interaction between *Cuscuta epithymum* and its host *Hormathophylla spinosa*, Gomez (1994) found that parasitized plants were visited by fewer pollinators than the non-parasitized plants and the diversity of the pollinator assemblage was also differed. Even the relationship between individual dodder and an individual host plant is neutral due to the array of direct and indirect positive and negative effects appearing between *C. epithymum* and *H. spinosa* during the successive phases of the reproductive cycle of the host plant. The exchange of chemicals between host and parasite was also documented and is evident based on a strong relationship between the parasite and its host, which should also be considered. Export of abscisic acid (ABA) from leaf petiole of *Pelargonium zonale* to *Cuscuta reflexa* shoot was reported by Bock and Fer (1992), while ABA was found to accumulate in the haustorium rapidly. The authors suggested that the very fast ABA uptake might has involved facilitated diffusion in addition to the diffusion of undissociated acid, and the high concentration of ABA in *C. reflexa*, especially in haustoria, could be explained by ABA import from host tissues. These authors concluded that ABA might have a central role in the host-parasite relationship by enhancing sucrose transfer. Kumar *et al.* (1993) studied *Cuscuta* sp. grown separately in pots and with host crops and indicated that the parasite did not require any stimulus for germination. Among the crops tested for host specificity, pulses in general were more seriously infested than oilseeds. The cereals were never infested by the parasite, which may be due to mechanical resistance. Baumel *et al.* (1994) reported the exchange of quinolizidine alkaloids between various host plants of the Fabaceae and parasitizing *Cuscuta* species.

Allelochemicals have been reported to play a significant role in the host/parasite interaction, at least during germination (Perez-de-Luque *et al.*, 2001). It is known that a number of host plants produce and excrete chemicals, which induce parasitic plant seed germination, while others inhibit or kill germinated seeds (Jorrin *et al.*, 1999). However, both parasites and host plants may release phytotoxic materials against each other. The former may use these chemicals to break down any defense mechanism arising by host plants or counteract the effect of host chemicals and facilitate dissolvent or separation of its cells and tissues and thus enhance parasite attachment and haustoria embedment in host organ. In contrast, host phytotoxic chemicals may be released to prevent attachment and act as repellents to the parasite or mask its ability to successfully attach itself to the host plant or prevent its development. Sinebo and Drennan (2001) suggested that sorghum released resistance-conferring substances to the infection points of *Striga hermonthica* after sensing infection. When infection points are widely distributed, in fully infected sorghum, fewer such substances appear to render the host more vulnerable. Mohamed *et al.* (2001) suggested the potential release of germination inhibitors from sorghum genotype P78 since *Striga* seeds near its roots consistently showed lower germination after ethylene treatment than those further out in the Petri-dish. Pigeon pea (*Cajanus cajan*), a trap species for *Orobanche* spp., was reported to affect different weed species through an allelopathy mechanism (Semidey and Bosques-Vega, 1999). In contrast, Bell-Lelong *et al.* (1994) suggested that *Striga* shows allelopathic activity on its host by means of a toxin and this was proved by the presence of high levels of oxidizable phenols and oxidative enzymes in *Striga hermonthica* shoot extracts. On the other hand, Perez-de-Luque *et al.* (2001) found that a susceptible variety of sunflower to *O. cumana* excreted a higher amount of secondary metabolites inducing parasite germination than the resistant one or more specific compounds.

Allelochemicals production is genetically controlled at which an allelochemical-control gene is involved. Yoder *et al.* (2001) reported that when parasitic plants are exposed to host root exudates, genetic pathways for both allelochemical detoxification and developmental signaling are invoked. A xenognostic organogenesis mechanism is invoked and parasitic weeds may have evolved the ability to use host allelochemicals against neighbouring plants. Mainly *Orobanche* and *Striga* depend on seed germination stimulants received in the root exudates of their hosts, suggesting an expression of several parasite genes (Atanasova *et al.*, 2001). *Striga* adaptation to parasitism also includes dependence on developmental signals and therefore, parasite and host development are highly integrated. It was reported that early host-derived chemical signals that *Striga* requires are mainly exuded from host roots into the soil (Butler, 1995).

Host plants were found to respond in different ways to parasite attack (Westwood *et al.*, 1996), among which lignifications for the parasite or host cell wall at the point of parasite penetration were reported by different authors (Arnaud *et al.*, 1998; Hood *et al.*, 1998). Among the defense responses used by plants are phytoalexins which can be attributed in part to altered regulation of the isoprenoid biosynthetic pathway (Griffitts *et al.*, 2001), leading to the production of many compounds necessary for general metabolism and growth.

In contrast, some parasitic species were reported to contain different forms of allelochemicals of various inhibitory properties. Dini *et al.* (1995) reported phenolic metabolites from dried and aerial parts of *Orobanche speciosa* that gave this species antimicrobial property. Saadoun *et al.* (2001) showed that spike tissues of different *Orobanche* species contain both phenols and mannitol and may act in the defense mechanism of these parasites against bacterial infection. These authors concluded that *Orobanche* tissues could be a potential source for antipathogenic bacterial agents. These allelochemicals may also be used against host plants to facilitate parasite invasion. However, the failure of host plants from defending themselves against parasites attack despite the induction of defense-related genes was in part addressed by Bell-Lelong *et al.* (1994) by that defense compounds produced may be degraded by the parasite or phytoalexins produced by the host could be not specifically toxic to the parasite.

Parasitic weeds are classified based on the host portion attacked into root parasites and stem parasites. Examples on root parasites are species of *Orobanche*, *Cistanche, Osyris, Thesium* and *Cynomorium.* Stem parasites include species of *Cuscuta*, *Loranthus* and *Viscum* that establish on the aerial parts of their hosts. Parasitic weeds are also described separately based on their botanical affiliation regardless of the plant portion attacked or the presence or absence of chlorophyll and feeding habit. Under botanical families and for each parasitic genus, the relevant literature was cited as deemed necessary.

This book provides a brief description of phenology, ecology, host range, spread and management of parasitic weeds of crop and forestry plants in Jordan. In addition, any possible economic or ecological value of these species through their use in traditional medicine based on their positive or negative attributes, their conservation or eradication systems that have been developed globally are also mentioned. It also aims to: (1) present parasitic flowering plants of cultivated crops and wild-grown species in Jordan; (2) familiarize agriculturists with the threat these species exert on agriculture; and (3) visualize the significant importance and damages they cause to agriculture and ecosystem realizing their distribution and hosts.

Parasitic Weeds of Jordan

(Part 1: Root Parasites; Orobanchaceae, Santalaceae & Cynomoriaceae)

GENERAL BACKGROUND

Weeds represent a major threat to agriculture in Jordan and cause great yield losses due to their negative interference with crop plants (Qasem, 2003). Weeds affect crop plants through competition, allelopathy, some through parasitism and may be in some cases through all at once. Parasitic weeds are one group of a deep concern to local farmers and researchers and exert a real danger to agriculture hence resources from water and nutrients are extremely limited considering that 90% of the country's area is a desert with drought and poor soil resources while the rest is devoted to forestry and agriculture. These parasites reported from the country (Table 1) subsist on the host species' root system or aerial vegetative parts. They can lead to severe growth damage hence obtain food, water, and nutrients from host plants, resulting in yield failure and death of the host plants under heavy infestation. Parasitic weeds may or may not have chlorophyll pigments and thus are partially or completely dependent on host plants for food and/or water (Parker and Riches, 1993). In Jordan, three families, including three genera, were reported to attack 34 wild and cultivated species (Abu-Iramileh, 1979). In more recent studies, Qasem (2009, 2011) reported eight parasitic genera of six plant families attacking many host species, among which 85 woody plant species from 57 genera belong to 33 plant families. The distribution of parasites throughout the country is shown in Fig. 3. The danger these parasites exert is mainly due to their difficulty to control with available weed control methods, except by soil fumigation mainly using methyl bromide, which has been banded a few years ago and its use is not economically feasible in the open field or for application at a wide scale. In addition, the huge number of seeds these parasites produce is accompanied by long seed viability and adaptability to disseminate by wind (*Orobanche*, *Cistanche*, and *Cynomorium*), animals or birds (*Loranthus*, *Viscum* and *Osyris*), or through crop infested seeds (*Cuscuta*, *Orobanche* and *Thesium*) add more complications to control methods. The interrelationship between the parasites and their hosts and how parasites connect to host plants through the absorptive organs (haustoria) is another means through which they challenge control measures. Morphological and physiological similarities between the parasitic species and their hosts and the ability of certain species to develop additional shoots from epicortical roots (*Loranthus* spp.) or embedded haustoria (*Cuscuta* spp.), especially after damage to or removal of the primary shoot (*Viscum* and *Loranthus*) hinder pruning and herbicides use as common methods of control.

Table 1. Parasitic plant species in Jordan.

Family	Genus	species	Reference
CUSCUTACEAE	*Cuscuta*	*Cuscuta approximate* Babingaton *Cuscuta babylionica* Choisy *Cuscuta brevistyla* A. Richard *Cuscuta campestris* Yuncker *Cuscuta epilinum* Weihe *Cuscuta gennesaretana* Sroelov *Cuscuta monogyna* Vahl *Cuscuta palaestina* Boiss. *Cuscuta planiflora* Ten. *Cuscuta pedicellata* Ledeb. *Cuscuta epithymum* (L.) L.	Musselman, 1998; Abu-Irmaileh, 1979; Qasem, 2008, 2010, 2011
CYNOMORIACEAE	*Cynomorium*	*Cynomorium coccineum* L.	Qasem, 2009, 2010, 2011,
LORANTHACEAE	*Loranthus*	*Loranthus acaciae* Zucc. (= *Plicosepalus acaciae*)	Abu-Irmaileh, 1979; Qasem, 2009
OROBANCHACEAE	*Cistanche* *Orobanche*	*Cistanche salsa* (C. A. Meyer) G. Beck *Cistanche tubulosa* (Schenk) Wight Syn. *C. lutea* *Orobanche aegyptiaca* Pers. *Orobanche cernua* Loefl *Orobanche crenata* Forskal *Orobanche grisebachii* Reuter *Orobanche lavandulacea* Reichenb *Orobanche minor* Sm *Orobanche mutelii* F. W. Schultz *Orobanche nana* Noe *Orobanche palaestina* Reuter *Orobanche ramosa* L. *Orobanche schultzii* Mutel	Qasem, 2006, 2009, 2010, 2011
SANTALACEAE	*Osyris* *Thesium*	*Osyris alba* L. *Thesium bergeri* Zucc. *Thesium humile* Vahl	Qasem, 2006, 2009, 2010, 2011, 2020
VISCACEAE	*Viscum*	*Viscum cruciatum* Siebr. ex Boiss.	Abu-Irmaileh, 1979; Qasem, 2009

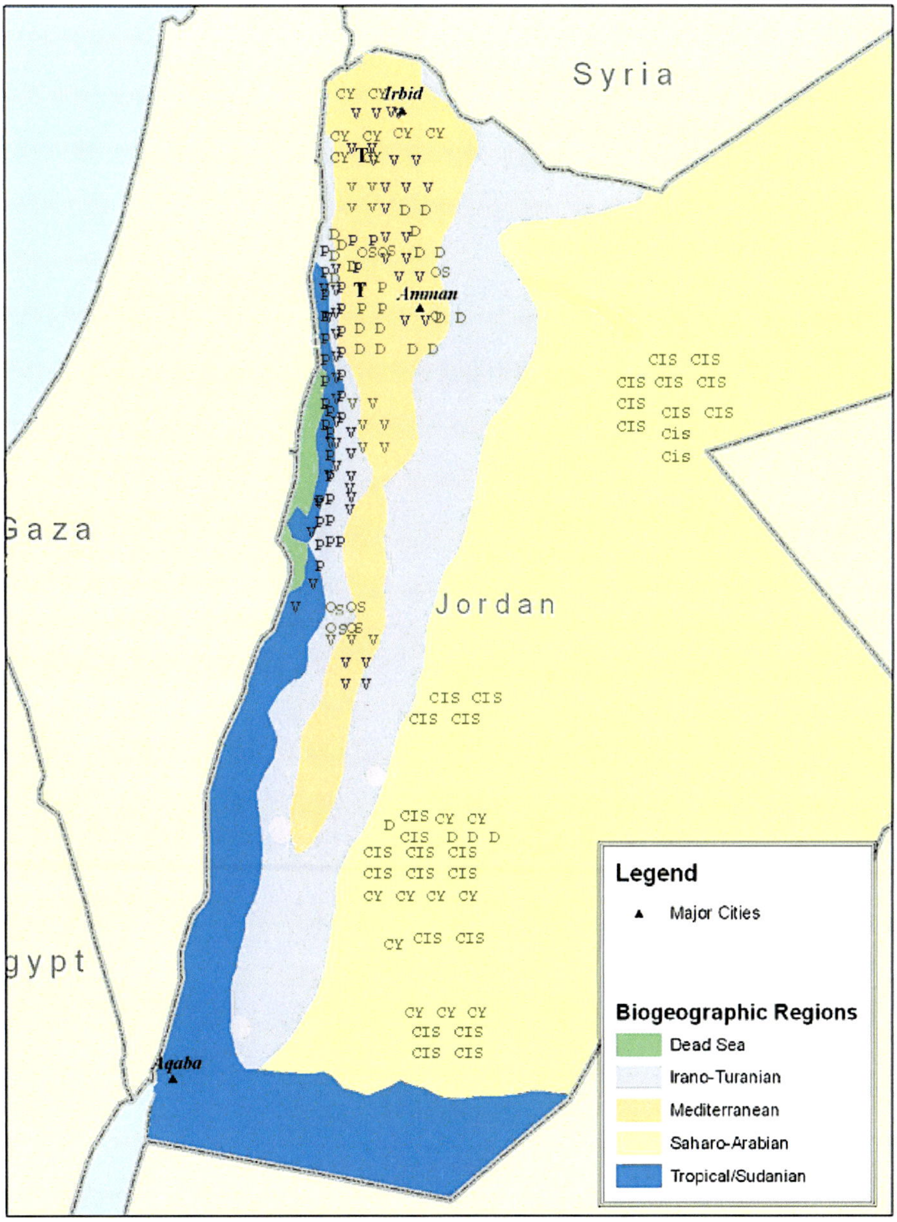

Fig. 3. Parasitic species distribution in the biogeographical regions of Jordan. V= *Viscum*, O = *Orobanche*, CIS = *Cistanche*, P = *Plicosepalus* (*Loranthus*), OS = *Osyris*, D = *Cuscuta*, CY = *Cynomorium*, T = *Thesium humile*.

PARASITIC PLANT FAMILIES IN JORDAN

1. Cuscutaceae: 6 species, attack 120 host species (41 cultivated) of 37 families

C. campestris 78 host species
C. monogyna 52 species
C. epithymum 14 species
C. epilinum 7 species
C. planiflora 5 species
C. indecora 2 species

Most parasitized families: Compositae (15 species), Leguminosae (14 species), Graminae (9 species) and Solanaceae (8 species)

2. Orobanchaceae:

a. Orobanche: 7 species, attack 86 host species (50 cultivated) of 24 families

O. ramosa 73 host species
O. cernua 20 species
O. aegyptiaca 12 species
O. crenata 7 species
O. palaestina 6 species
O. minor 2 species
O. schultzii 2 species

Most parasitized families: Compositae (20 species), Solanaceae (11 species), Leguminosae (9 species), Umbelliferae (7 species), Cruciferae (7 species), Cucurbitaceae (4 species), Labiatae (4 species) and Rosaceae (4 species)

b. Cistanche: 3 species, attack 20 host species of 7 families

C. tubulosa 19 host species (connection with 5 species was not checked)
C. lutea 1 species
C. salsa 10 species (connection with 2 species was not checked)

Most parasitized families: Chenopodiaceae (7 species), Leguminosae (3 species)

3. Loranthaceae

Loranthus (Plicosepalus) acaciae: attacks 26 plant species of 12 plant families

Most parasitized families: Leguminosae (11 species), Anacardiaceae (3 species) and Rhamnaceae (3 species)

4. Viscaceae

Viscum cruciatum: attacks 14 plant species of 8 plant families

Most parasitized families: Rosaceae (6 species) and Fabaceae (2 species)

5. Santalaceae

a. Osyris alba: 23 host species (connection with 5 species was not checked) of 15 plant families

Most parasitized families: Leguminosae (4 species), Rosaceae (4 species), Rhamnaceae (2 species)

b. Thesium humile: 2 host species

Most parasitized families: Liliaceae (1 species) and Gramineae (1 species)

6. Cynomoriaceae

Cynomorium coccineum: attacks 5 plant species

Parasitized families: Chenopodiaceae (3 species), Labiatae (1 species) and Umbelliferae (1 species).

MANAGEMENTS OF PARASITIC WEEDS IN JORDAN

Control methods of parasitic flowering plants in Jordan are generally traditional or absent for certain parasitic genera. However, farmers' practices in controlling these parasites are shown in Table 2. Hand removal is widely practised in different locations and for different parasites. A combination of more than one control method is commonly followed against certain genera (e.g. *Orobanche* and *Cuscuta*), while hand removal and pruning are both practised for mistletoes control. In contrast, hand pulling is the only operation practised against *C. coccineum,* while farmers are not familiar with *Osyris alba* as a parasitic species. Application of soil solarization followed by plastic mulching of the solarized soil and application of chicken manure and\or ammonium sulphate or urea fertilizers are common practices in the Jordan Valley and may be in other regions. These practices have effectively restricted the infestation of *Orobanche* spp. and to a lesser extent of *Cuscuta* infestation in vegetable crops. Soil plastic mulching can also be applied in certain fruit tree orchards. This, however, is under experimentation at present while the application of animal manure and straw mulch is commonly practiced in different regions, which may lower infestation by certain root parasites (mainly *Orobanche*). Control of other parasitic species is still in infancy. Mistletoes are usually pruned at late stages and hand removed at early infestation. However, the second practice is not effective in most cases since parasites can re-vegetate from the inserted suckers inside the host tissues and are difficult to adopt against *Loranthus* because many epicortical roots are extended on the host stem surface. Herbicide application for thecontrol of these parasites is completely absent. *Osyris alba* control is not practised by any means except through removal by tillage or the hoe without any previous knowledge of the farmers on this parasite. Control of *Cynomorium* and *Cistanche* spp. is mainly practised through a publiccollection of both parasites in different regions, while farmers are not concerned enough about losses resulting from these parasites since they mostly attack forage plants and wild shrubs in the desert. Natural enemies of some parasites may play an important role in their control, although no studies on these were conducted under local conditions. However, some of the parasites are attacked by different insects. The larvae of *Thrips* sp. was found in seed capsule (*C. tubulosa*), in the stem of *O. schultzii*, or on the leaves and branches of *V. cruciatum* (*Ceroplastes rusci*). Al-Khesraji *et al.* (1987) reported that different natural enemies attacked *Cynomorium* and *Cistanche* spp. in the southern desert of Iraq, including *Tropinota squalida* that attacking species of both genera, *Aphis gossypii* on *Cistanche* spp., larvae of *Eumerus* spp. and those of Lepidoptera (Fam. Sesiidae) on both genera and the larvae were found very effective on *Cynomorium. Phytomyza orobanchia* Kaltenbach (Diptera: Agrornyzidae) was also found in the seed capsules of certain *Orobanche* species. In addition, the same parasitic species are attacked by different fungi.

Table 2. Managements of parasitic weed control by local farmers in Jordan (Qasem, 2010).

Method of control	*Cuscuta* spp.	*C. coccineum*	*O. alba*	*V. cruciatum*	*L. acaciae*	*Cistanche* spp.	*Orobanche* spp.
Hand pulling/removal	+	+	-	+	+	+	+
Hoeing	+	-	-	-	-	+	+
Soil solarization	+	-	-	-	-	-	+
Planting date	+	-	-	-	-	-	+
Rootstocks/Grafting	-	-	-	-	-	-	+
Plastic mulch	+	-	-	-	-	-	+
Fertilizers (NH2SO4, Urea	-	-	-	-	-	-	+
Chicken manure	+	-	-	-	-	-	+
Intercropping	-	-	-	-	-	-	+
Pruning	+	-	+	+	+	-	-
Soil applied herbicides	≈	-	-	-	-	-	≈
Foliage applied herbicides	-	-	-	-	-	-	-

+ = Effective, - = Ineffective, ≈ = somehow effective but not widely followed

REFERENCES

Abu-Iramileh, B.E. Occurrence of parasitic flowering plants in Jordan. *Plant Dis. Reptr.*, 1979, 63, 1025-1028.

Albrecht, H., Yoder, J.I., Phillips D.A. Flavonoids promote haustoria formation in the root parasite *Triphysaria versicolor. Plant Physiol.* 1999. 119, 585– 591. doi:10.1104/pp.119.2.585.

Al-Khesraji, T.O., Abdel Wahib, A.U., Annou, M.R. Insect pests attacking parasitic flowering plants in the southern desert of Iraq. *Iraqi J. Agr. Sci. "Zanco"*, 1987,5, 197–216.

Aly, R. Conventional and biotechnological approaches for control of parasitic weeds. *In VitroCell. Dev. Biol Plant.*, 2007, 43, 304–317. DOI 10.1007/s11627-007-9054-5

Arnaud, M.C., Thalouarn, P., Fer, A. Characterization of mechanisms related to the resistance of crops to two parasitic higher plants (*Cuscuta reflexa* and *Striga hermonthica*). *Comptes Rendus des Seances de la Societe de Biologie et de ses Filiales.*, 1998, 192, 101-119.

Atanasova, S., Hoefsloot, G., Verkleij, J.A.C. T-DNA tagging in *Arabidopsis*: An application for *Orobanche* research. In: Fer, A., Thalouarn, P., Joel, D.M., Musselman, L.J., Parker, C., Verkleij, J.A.C., (eds.), *Proceedings of the 7th International Parasitic Weed Symposium,* Nantes, France, 2001, pp. 92-95.

Baumel, P., Witte, L., Czygan, F.C., Proksch, P. Transfer of quinolizidine alkaloids from various host plants of the Fabaceae to parasitizing *Cuscuta* species. *Biochemical Systematics and Ecol.*, 1994, 22, 647-656.

Bell-Lelong, D., Butler, L.G., Ejeta, G., Hess, D. Do phenolics from the parasitic weed *Striga* inhibit host growth? Geibel, M., Treutter, D., Feucht, W., (eds.). *Acta Hortic.* 1994, 381(II), 683-686.

Bharathalakshmi, S., Jayachandra, A. Studies on the germination of some Indian collections of *Striga asiatica* (L.) Kuntze. In: Parker, C., Musselman, L.J., Polhill, R. M., Wilson, A.K., (eds.), *Proceedings, 3rd International Symposium on Parasitic Weeds*, Aleppo, Syria, ICARDA, Aleppo, 1984, pp. , 108-117.

Bharathalakshmi, S., Werth, C.R., Musselman L.J. A study of genetic diversity among host specific populations of witchweed *Striga asiatica* (Scrophulariaceae) in Africa. *Plant Syst. Evol.*, 1990, 172, 1-12.

Bock, F.de, Fer, A. Effects of abscisic acid on the transfer of sucrose from host, *Pelargonium zonale* (L.) Aiton, to a phanerogamic parasite, *Cuscuta reflexa* Roxb. *Aust. J. Plant Physiol.*, 1992, 19, 679-691.

Butler, L. G. Chemical communication between the parasitic weed *Striga* and its crop host. A new dimension in allelochemistry. In: Inderjit, Dakshini, K.M.M., Einhellig, F.A., (eds.), *Allelopathy: Organisms, Processes, and Applications*, 1995, pp. 158-168.

Cudney, D.W., Orloff, S.B., Reints, T.S. An integrated weed management procedure for the control of dodder (*Cuscuta indecora*) in alfalfa (*Medicago sativa*). *Weed Technol.*, 1992, 6, 603-606.

Dhanapal, G.N. *Management of Broomrape (Orobanche cernua) in Tobacco* (Nicotiana tabaccum). Dissertation, Department of Agronomy, Wageningen Agricultural University, Wageningen, Netherlands, 1996, 183 pp.

Dini, I., Iodice, C., Ramundo E. Phenolic metabolites from *Orobanche speciosa. Planta Med.*, 1995, 61(4), 389-390.

Fernández-Aparicio, M., Yoneyama, K. Rubiales, D. The role of strigolactones in host specificity of *Orobanche* and *Phelipanche* seed germination. *Seed Sci. Res.*, 2011, 21, 55–61.

Galindo, J.C.G., Macías F.A., García-Díaz, M.D., Jorrín, J. Chemistry of Host-Parasite Interactions. In: Macías, F.A., Galindo, J.C.G., Molinillo J.M.G., Cutler, H.G. (eds.),

Allelopathy. Chemistry and Mode of Action of Allelochemicals., CRC Press, Boca Raton, FL, 2004, pp. 125-148.

Gomez, J.M. Importance of direct and indirect effects in the interaction between a parasitic angiosperm (*Cuscuta epithymum*) and its host plant (*Hormathophylla spinosa*). *Oikos*, 1994, 71, 97-106.

Griffitts, A.A., Gramer, C.L., Westwood, J.H. Characterization of root plant responses to parasitization by *Orobanche aegyptiaca*. In: Fer, A., Thalouarn, P., Joel, D.M., Musselman, L.J., Parker, C., Verkleij, J.A.C., (eds.), Proceedings of the 7th International Parasitic Weed Symposium, Nantes, France, 2001, 178-181.

Hauck, C., Müller, S., Schildknecht, H. A germination stimulant for parasitic flowering plants from *Sorghum bicolor*, a genuine host plant. *J. Plant Physiol.*, 1992, 139, 474–478.

Heide-Jorgensen, H.S. The Parasitic syndrome in higher plants. In: Joel, D.M., Gressel, J., Musselman, L.J., (eds.), *Parasitic Orobanchaceae*. Springer- Verlag Berlin Heidelberg, Germany, 2013, pp. 1-18.

Hood, M.E., Condon, J.M., Timko, M.P., Riopel, J.L. Primary haustorial development of *Striga asiatica* on host and nonhost species. *Phytopathology*, 1998, 88, 70-75.

Hudu, A.I., Gworgwor, N.A. Preliminary results on evaluation of trap crops for *Striga hermonthica* (Del.) Benth. control in sorghum. *International Sorghum and Millets Newsletter*, 1998, 39, 118-121.

International Allelopathy Society (IAS). Constitution and Bylaws. First World Congress on Allelopathy: A science for the future. University of Gadiz, Gadiz, Spain. 1996.

Joel, D., Goldman-Guez, T. The mode of *Cuscuta* haustorial penetration. In: Fer, A., Thalouarn, P., Joel, D.M., Musselman, L.J., Parker, C., Verkleij, J.A.C., (eds.), *Proceedings of the 7th International Parasitic Weed Symposium,* Nantes, France, 2001, 32-35.

Jorrin J., Perez-de-Luque A., Serghini K. How plants defend themselves against root parasitic angiosperms: Molecular studies with *Orobanche* spp. In: Cubero, J.I., Moreno, M.T. Rubiales; Sillero, D. (eds.), *Resistance to Orobanche: The state of art*, Cordoba, Spain, 1999, 163-177. Sevilla (Espana): JA, DGIFA, 1999, ISBN 84-89802-51-3. 199 p.

Kumar, R.M., Kondap, S.M., Madhav Rao, N., Ramakrishna, A. An anatomical study on host specificity of (a) parasitic weed-*Cuscuta*, integrated weed management for sustainable agriculture. In: Indian Society of Weed Science, *Proceedings of an Indian Society of Weed Science International Symposium*, Hisar, India, 18-20 November 1993, II, Hisar, Haryana, India: Indian Society of Weed Science pp. 61-63.

Machado, M.A., Zetsche, K.A structural, functional and molecular analysis of plastids of the holoparasites *Cuscuta reflexa* and *Cuscuta europaea*. *Planta*. 1990, 181, 91-96.

Mallet, A.I. Studies in the chemistry of the *Orobanche crenata* germination factor present in the roots of *Vicia faba* and other host plants. In: EWRC, *Proceedings, Symposium on Parasitic Weeds, Malta, 1973. European Weed Research Council*, Wageningen, 1973, pp., 89-97.

Mohamed, A., Rich, P.J., Housley, T.L., Ejeta, G. *In Vitro* techniques for studying mechanisms of *Striga* resistance in sorghum. In: Fer, A., Thalouarn, P., Joel, D.M., Musselman, L.J., Parker, C., Verkleij, J.A.C., (eds.), *Proceedings of the 7th International Parasitic Weed Symposium,* Nantes, France, 2001a, pp. 96-99.

Mohamed, A.H., Ejeta, G., Housley, T.L. *Striga asiatica* seed conditioning and 1-aminocyclopropane-1- carboxylate oxidase activity. *Weed Res.*, 2001b, 41, 165-176.

Muller, S, Hauck, C., Schildknecht, H. Germination stimulants produced by *Vigna unguiculata* Walp cv. Saunders. *J. Plant Growth Reg.*, 1992, 11(2), 77-84.

Musselman, L. J. (Ed.). *Parasitic Weeds in Agriculture*. Vol. 1. *Striga*. CRC Press. Boca Raton. 1987.

Musselman, L.J. *Checklist of Plants of the Hashemite Kingdom of Jordan.* Old Dominion University, 1998. Available at: http://www.odu.edu/~lmusselm/plant/jordan/index.php

Musselman, L.J. Parasitic weeds of the arable land. In: Holzner, W., Numata, M., (eds.), *Biology and Ecology of Weeds*, The Hague, Netherlands, Dr. W. Junk. 1982, 175-185.

Nickrent, D. L. Parasitic angiosperms: How often and how many? *Taxon.*, 2020. 69 (1), 5-27. Available at: https://doi.org/10.1002/tax.12195.

Nickrent, D. L. Plantas Parásitas en el Mundo. In: López-Sáez, J. A., Catalán, P., Sáez, L., Mundi-Prensa Libros, S. A., (eds.), *Plantas Parásitas de la Península Ibérica e Islas Baleares,* Madrid. 2002, Capitulo 2, pp. 7-27.

Parker C., Riches C.R. *Parasitic Weeds of the World, Biology and Control.* CAB International, Wallingford, UK. 1993.

Parker, C. Parasitic weeds: a world challenge. *Weed Sci.*, 2012, 60, 269–276.

Perez-de-Luque, A., Rubiales, D., Galindo, G.C., Macias, F.A., Jorrin, J. Allelopathy and allelochemicals within the plant-parasitic weed interaction. Studies with the sunflower-*Orobanche cumana* system. In: Fer, A., Thalouarn, P., Joel, D.M., Musselman, L.J., Parker, C., Verkleij, J.A.C., (eds.), *Proceedings of the 7th International Parasitic Weed Symposium,* Nantes, France, 2001 pp.196-199.

Plakhine, D., Ziadna, H., Joel, D.M. Is seed conditioning essential for *Orobanche* germination? *Pest Manag Sci.* 2009, 65, 492–496 (www.interscience.wiley.com) DOI 10.1002/ps.1716

Qasem, J. R. Parasitic weeds and allelopathy, from the hypothesis to the proof. In: Reigosa M.J., Pedrol, N., Gonzalez, L., (eds.), *Allelopathy, a Physiological Process with Ecological Implications*, Springer, the Netherlands, 2006, pp. 565- 637.

Qasem, J. R. Parasitic weeds of the Orobanchaceae family and their natural hosts in Jordan. *Weed Biol. Manag.* , 2009, 9 (2), 112-122.

Qasem, J. R. *Weeds and Their Control.* University of Jordan Publications. Amman, Jordan. 2003, 628 pp. ISBN 978-9957-420-05-5. 2003.

Qasem, J.R. (2020). Osyris (*Osyris Alba* L) Updates, the illustrated threatening parasite to fruit and forestry trees in Jordan. *Pak. J. Bot.*, 2020, 52(1), 251-256. DOI:10.30848/PJB2020-1(21)

Qasem, J.R. Dodder (*Cuscuta* spp.) Occurrence and natural hosts in Jordan. *Kor. J. Weed Sci.*, 2008, 28 (4), 343-359.

Qasem, J.R. Parasitic flowering plants of cultivated plants in Jordan, the present status and possible management. *Pak. J. Weed Sci. Res.*, 2010,16(2), 227-239.

Qasem, J.R. Parasitic flowering plants of woody species in Jordan. *Eur. J. Plant Pathol.*, 2011, 131, 143–155, DOI 10.1007/s10658-01-9794-2.

Qasem, J.R., Kasrawi, M.A.Variation of resistance to broomrape (*Orobanche ramosa*) in tomatoes. *Euphytica*, 1995, 81, 109-114.

Rice, E.L. *Allelopathy.* 2nd ed. Academic Press. Orlando. Fl. 1984, pp. 422.

Riopel, J.L., Timko, M.P. Signals and regulation in the development of *Striga* and other parasitic angiosperms. In: *Molecular Signals in Plant-Microbe Communications.* D.P.S. Verma. Boca Raton, CRC Press Inc., 1991, pp. 493-507.

Saadoun, I., Hameed, K.M., Al-Momani, F., Foy C.L. Effect of three *Orobanche* spp. on some local phytopathogens, *Agrobacterium* and *Erwinia*. In: Fer, A., Thalouarn, P., Joel, D.M., Musselman, L.J., Parker, C., Verkleij, J.A.C., (eds.), *Proceedings of the 7th International Parasitic Weed Symposium,* Nantes, France, 2001, P. 41.

Salle, G., Aber, M., Le-Coz, S. Comparative histo-physiological study of the germination and establishment of *Cuscuta*. In: Parker, C., Musselaman, L.J., Polhill, R.M., Wilson, A.K. (eds.), *Proceedings of the Third International Symposium on Parasitic Weeds*, Aleppo, Syria, ICARDA/1PSPRG, 7-9 May 1984. ICARDA, Aleppo, Syria, pp., 10-21.

Sauerborn, J. Angiospermous root parasites and fungi as natural antagonists. In: Fer, A., Thalouarn, P., Joel, D.M., Musselman, L.J., Parker, C., Verkleij, J.A.C., (eds.), *Proceedings of the 7th International Parasitic Weed Symposium,* Nantes, France, 2001, pp. 248-251.

Semidey, N., Bosques-Vega, A.Yield and weed suppression by pigeon pea cultivars in rotation with tomato and pepper. *Journal of Agriculture of the University of Puerto Rico*. 1999, 83(1/2), 55-64.

Siame, B.A., Weerasuriya, Y., Wood, K., Ejeta, G., Butler, L.G. Isolation of strigol, a germination stimulant for *Striga asiatica*, from host plants. *J. Agric. Food Chem.*, 1993, 41(9), 1486-1491.

Sinebo, W., Drennan, D.S.H. Vegetative growth of sorghum and *Striga hermonthica* in response to nitrogen and the degree of host root infection. *Eur. J. Plant Pathol.*, 2001, 107, 849-860.

Stewart, G. R., Press, M. C. The physiology and biochemistry of parasitic angiosperms. *Ann. Rev. Plant Physiol. Plant Mol. Biol.*, 1990, 41,127-151.

Subhashini, K., Kumar, P.K.R., Gaddeyya, G. A comprehensive review on *Dendrophthoe falcata* (L.f.) Ettingsh. (Loranthaceae). *Trop. Plant Res.,* 2019,6(3), 514–520.

Westwood, J.H., Yoder, J.I., Timko, M.P., de Pamphilis, C. W. The evolution of parasitism in plants. *Trends Plant Sci.*, 2010, 15(4), 227-235. doi: 10.1016/j. tplants.2010.01.004

Westwood, J.H., Yu, X., Foy, C.L., Cramer, C.L. Parasitization by *Orobanche* induces expression of a defense-related gene in tobacco. In: Moreno, M.T., Cubero, J.I., Derner, D., Joel, D., Musselman, L.J., Parker, C., (eds.), *Advances in Parasitic Plant Research, Sixth International Parasitic Weed Symposium*, April 16-18, 1996, Cordoba, Spain, 1996, pp. 543-550.

Whitney, P.J. Broomrape (*Orobanche*) seed germination inhibitors from plant roots. *Ann Appl Biol*, 1978, 89, 475-478.

Yoder, J.I., Matvienko, M., Jamison, D., Torres, M., Ngo, Q., Nava, A.J., Nava, A., Goldwasser, Y., Wojtowicz, A., Wrobel, R. Early parasite responses to hosts root factors. In: Fer, A., Thalouarn, P., Joel, D.M., Musselman, L.J., Parker, C., Verkleij, J.A.C., (eds.), *Proceedings of the 7th International Parasitic Weed Symposium*, Nantes, France, 2001, pp. 5-10.

Yokota, T., Sakai, H., Okuno, K., Yoneyama, K., Takeuchi, Y. Alectrol and orobanchol, germination stimulants for *Orobanche minor*, from its host red clover. *Phytochemistry*, 1998,49, 1967-1973.

SECTION I

THE FAMILY: OROBANCHACEAE

CHAPTER 1

Broomrapes (*Orobanche* spp.)

Jamal Ragheb Said Qasem

INTRODUCTION

Orobanchaceae is mainly a northern warm and temperate zones family, around 10% of the species belong to this family are tropical, and only one species reaches the arctic (Eltayeb *et al.*, 2000). Reports on the number of genera and species included in this family and their distribution are quite varied. The family was mentioned to include 15 genera and more than 200 species (Kabulov, 1978); it is of 15 obligate parasitic genera with 250 species (Pusch and Günther, 2009), presents the largest parasitic family of 20 genera of root holoparasites and 270 species spread in all climatic zones of the globe excluding Antarctica. The family includes 85 genera consist 1650 species (Nickrent and Musselman, 2004). More recent studies supported by molecular taxonomy analyses have resulted in re-definition of the family as consists 89 genera, including 2061 species (Nickrent, 2008; Gevezova *et al.*, 2012).

The majority of species are facultative or obligate root parasites that subsist on broad-leaf plants and exhausting hosts' nutrients, minerals and water (Gevezova *et al.*, 2012). The main and representative genus of the family is *Orobanche*, regarded as the most important and destructive root parasite. *Orobanche* includes more than 160 angiosperm parasitic species (Oliveira Velloso, 1990); the majority (about 90%) are of the old world. The most crop destructive and problematic species of *Orobanche* are branched broomrape (*O. ramosa*), Egyptian broomrape (*O. aegyptiaca*), nodding broomrape (*O. cernua*), crenate broomrape (*O. crenata*), and small broomrape (*O. minor*); some workers also added the sunflower broomrape (*O. cumana*) and fetid broomrape (*O. foetida*) as economically important (Parker, 2009). The main hosts of these parasites are species of Leguminosae, Apiaceae and Asteraceae for *O. crenata*, the sunflower (*Helianthus annuus*) for *O. cumana*, Solanaceae for *O. cernua*, Leguminosae for *O. foetida*, clover (*Trifolium* spp.) and alfalfa (*Medicago sativa*) for *O. minor*, and many families, including the Solanaceae, Cucurbitaceae, and Brassicaceae for *O. ramosa* and *O. aegyptiaca* (Qasem, 2009; Parker, 2012; Punia, 2014). However, different *Orobanche* species may also attack species of different plant families. For instance, *Orobanche minor* was reported to parasitize a diverse range of flowering plants from at least 16 orders of monocots and dicots (Thorogood *et al.*, 2009). This species is of a widespread in southeastern England and of only transient occurrences on disturbed sites in Scotland while the commonest variant, var. *minor*, may well be an ancient introduction from Central Europe (Silverside, 2002a, b).

Orobanche species are temperate and often highly selective in their habitats. Usually prefer dry, open, nutrient-poor soils, where they are destructive to agricultural and scrub development or other types of habitat loss (Parker and Riches, 1993; Qasem, 2006b). Some also appear sensitive to minor climatic changes and may disappear from sites that still appear suitable. Many are extremely local throughout their entire geographical regions. *Orobanche* spp. are flowering holo-root-parasites lack chlorophyll and therefore are not able to produce their own food (heterotrophic), while leaves are vestigial. Parasite shoot/s appears above the soil almost at the end of the parasites life cycles and only for flowering and seed production (Nickrent and Musselman, 2004; Fernández-Aparicio *et al.*, 2016).

Generally, *Orobanche* attacks roots of broadleaf plants from which they suck nutrients and water, and complete their life cycle within a year, hence considered annuals in duration. They are prolific seed producers, varied in the number of seeds produced and seed longevity period in the soil. Seeds for all species are tiny (Fig. 1.1.1) measuring 0.2 by 0.3mm, with minute amount of reserved food and differences exist in flower and seed size between species and in flower colour and other morphological variations.

Fig. 1.1.1. *Orobanche ramosa* seeds

Species bear seeds in some huge numbers and of near-microscopic size (*e.g. O. minor* and *O. ramosa*). For example, 1000 seeds of *O. elatior* weigh 0.0049 g and one plant produces 270.000 seeds. *Orobanche picridis* produces from 94000 to 116000 seeds per plant and 1000 seeds weigh 0.0029 g. Seeds are widely disseminate by wind and, though minute, are very long-lived, having been known to survive in the soil for 20 years, and germinating in succession but under low moisture level dormancy may be extended up to 30 years. *Orobanche* species are also different in branching habit and in presence or absence of bracteoles, characters that distinguish species and regarded as useful tools in identification. *Orobanche ramosa* and *O. aegyptiaca* are branched and bracteoles, *O. cernua*, *O. minor* and *O. crenata* are usually of a single unbranched stem and lack bracteoles. However, differences between species of each group in stature, flower size and color and host range are evident (Parker and Riches, 1993). More differences between species were also detected in distribution, parasite forms or variations and losses caused to crop plants.

GERMINATION AND DEVELOPMENT

Orobanche seeds are of minute amount of stored food, generally remain viable in soil for 10 to 13 years (Brenchley, 1920) or up to 20 years (Puzzilli, 1983). Seeds in general and of certain species (*e.g. crenata*) in particular are dormant and must undergo a period of post-harvest or after ripening which may extend from 18-24 months in warm, dry storage, followed by conditioning period in a warm, moist environment, before respond to germination stimulants. The length and effectiveness of the after ripening period may vary with temperature, humidity, and other environmental factors. The nature of the changes occur in the seeds during after ripening are unknown but various structural and metabolic changes in the seed coat and embryo may occur, and changes in the respiratory substrate are also involved. Mature seeds require conditioning period between 1-3 weeks although the duration is dependent on certain factors like temperature, origin, and age of the seeds but ranges from several days up to several weeks depending on the species during which seed coat permeability may increase, water absorption occurs and/or may be changes in the levels or activities of endogenous germination promoters or inhibitors. Some chemicals such as fluridone can shorten the conditioning period required for *Orobanche* spp. seeds (Song *et al.*, 2005) in response to a germination stimulant. It can also enhance the response of *Orobanche* spp. seeds to the artificial strigol-analogue germination stimulant GR24, even when the seeds are first conditioned at a suboptimal temperature and under water stress (Song *et al.*, 2006).

Conditioning temperature varied between species but range between 18-25°C however, fluctuation between conditioning and germination temperature enhanced germination. Seeds germinate when are close to or in contact with host roots, triggered by chemical recognition, hence must receive germination stimulants in root exudates of host species while host roots should be few mm from germinated seeds. In general germination occurs in the immediate vicinity of the host root. It appears that the chemical stimulant secreted by the host root system can cause germination up to a distance of 1 cm from the root but only seeds within 2-3 mm of the root infest the host. Optimal temperature for germination is between 20-25°C, but the lower temperature, the slower is the *Orobanche* development. At 30/20°C a cardinal temperature is reached wheredevelopment is also delayed. *Orobanche crenata* prefers to germinate at 15/5°C, while optimum temperature for attachment is 20/10°C. Other species germinate in a broader range of temperature (10/5 to 25/15°C). *Orobanche crenata* mainly attacks winter crops, while *O. aepyptiaca* parasitizes summer and winter species and *O. cumana* prefers summer crops. Poor germination at low and high temperatures may be due to the lack or shortage in host root exudates, inactive stimulants or failure of seeds to germinate under these conditions. Generally, *Orobanche* germination occurred at as low as 5°C but is very much delayed. For *O. crenata* and *O. cumana* eventhough attachment is observed at this low temperature but was not for *O. aegyptiaca*. Other edaphic and climatic conditions influence parasite seed germination for instance soil alkalinity tends to inactivate germination stimulants and thus reducing parasite infestation while conflict on the effect of light during conditioning or germination phases of *Orobanche* seeds were reported and differences between parasite species in their responses to this factor also have been documented. Light enhanced *O. ramosa* germination but inhibited seed germination of *O. crenata*.On the other hand, change in the pH or presence of calcium or phosphate ions that modify the effect of pH had a significant effect on germination stimulants while inactivation of germination stimulants in the soil was found to be due to microorganisms and the effect was varied in different soil types. Exudates concentration is another factor affects parasite seed germination at which high concentration reduces germination due to components rather than stimulants. It seems that plant exudates contain more than one stimulant at which some trigger particular species to germinate butnot others. These may explain the variation in host specificity of different *Orobanche* species and differences in the degree of specificity shown by seeds of the same species but of different origin. Different materials were found to enhance seed germination of *Orobanche* species among which nonhost root exudates of flax (*Linum usitatissimum*), maize (*Zea mays*), cotton (*Gossypium hirsutum*), chili and bell peppers (*Capsicum annuum* L.), coat buttons (*Tridax procumbens* L.), andhairy beggar-ticks (*Bidens pilosa* L.) (Qasem, 2006a; 2019).When broomrape seed occurs close to the roots of such species, 13 to 15% of seeds germinate but failed to develop haustorial connections. Ethylene gas was reported to enhance seed germination of *O. ramosa* but conflict reports are also available on its effect. Among other chemicals reported to enhance seed germination are gibberellic acid and calcium hypochlorite. In contrast, phosphate fertilizer decreased *Orobanche* incidence.

CONTACT AND ATTACHMENT

Germination of *Orobanche* seeds is confirmed by the extension of a radicle (germination tube) that capable to grow for 1-2mm before the seeds' stored food is exhausted. The radicle moves toward the host root system in a chemotropic response or bending movement in response to inhibitory chemical substances in the root exudates of host plants. For successful contact, the parasite radical must find the host root, then development of the haustoria should take place and finally penetrate into the host root through the epidermis and cortex and connection between vascular systems of the parasite and the host plant be developed. This type of germination is probably highly evolved and acts to enhance seedling survival. Upon the contact between the parasite and its host, sticky papillae are formed, which may adhere the parasite to the root surface. Penetration of the developing haustoria into host root tissues occurs mainly by mechanical pressure on the host endodermal cells and enzymatic action through hydrolytic enzymes. Radicle growth of the germinated seed of the root parasite *O. ramosa* is shown to be rapidly accompanied by secretion of proteins including pectinolytic enzymes, polygalacturonase and rhamnogalacturonase (Veronesi *et al.*, 2007). The symbiotic relationship between microorganisms and the host root system may enhance penetration of *Orobanche* haustoria through rhizobium nodules on the roots of *Vicia faba* L. or may be other legumes. The primary haustoria and the tubercles are the major connecting organs (Fig. 1.1.2.). When the attachment occurred, the germination tube pulls the active meristem of the embryo from the seed to the root surface of the host plant and the radicle doesn't act anymore. The development of the haustoria takes place and the haustoria swell forming tubercles or nodules (Fig. 1.1.3) and establish connections with the vascular system of host plants (Dorr and Kollmann, 1975; 1976). In general, parasite germination, attachment and haustorial formation and penetration are mediated by chemical communication between the host and the parasite. Once the connection is established between the host and the parasite vascular systems, the latter can start absorbing the required materials, growing and developing. The broomrapes are phloem-feeding parasites that absorbed nutrients predominantly from the phloem tissues of the host plant (Irving and Cameron, 2009); although phloem-feeding parasites keep a xylem connection but obtain most of the C and N required from the phloem of their hosts and are often classified as obligate holo-parasites. The attached parasite functions as a powerful supersink that strongly deprives water, mineral nutrients and carbohydrates absorption and translocation through the haustorial bridge connection between the parasite and host, causing drought stress and wilting of the host plant. *Orobanche* accumulates high amounts of mannitol that lowers its osmotic potential to a much more negative value than the host (Harloff and Wegmann, 1987). This osmoregulation process, essential to broomrape for water and nutrients uptake from the host, is absent from the major host crops. Upon haustorial growth, an underground tuber develops, from which the shoot bud differentiates, elongate and emerges above the soil for flowering and seeding.

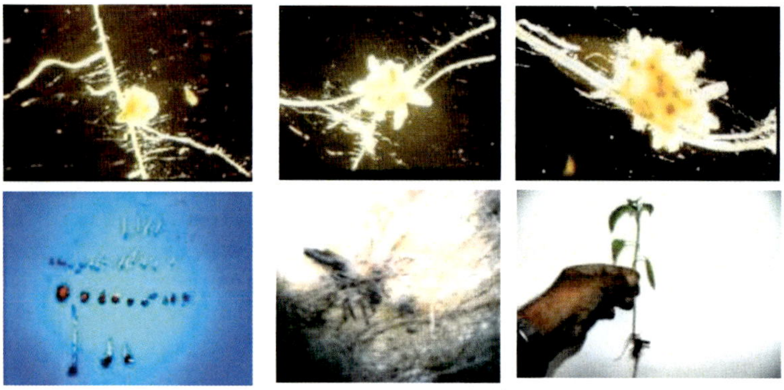

Fig. 1.1.2. Haustoria developmental stages

Orobanche seedlings are highly sensitive, light in color, almost transparent, rootless and devoid of chlorophyll. Although it easily penetrate the soil, but can't absorb or directly draw soil growth factors. However, in the absence of a suitable host soon after seed germination, the seedlings shrivel and die.

Orobanche life cycle consists of two phases; the first is the preconditioning during which the seed embryo conditioning takes place under suitable temperature and moisture. This period was considered important for leaching germination inhibitors from parasite seeds (Edwards, 1972); while other workers (Jain and Foy, 1992; Ibrahim *et al.*, 2000) suggested that germination stimulants or related substances accumulate during this period. The second phase is seedlings' attachment to the host root system and the start of true parasitism. This phase can be divided into two stages, the hypogeal stage that occurs below the soil level and lasting for 60 days (Mukumov, 1974), during which the vegetative subterranean organ develops and food materials absorbed, and the epigeal stage at which the vegetative organ grow rapidly and produce reproductive organs above the soil. The emergence, flowering and seeding of the parasite may last for only 15 to 18 days (Mukumov, 1974). However, the hypogeal stage is responsible on most of the damage that parasite cause to the host plant, while the epigeal stage occurs mainly for flowering and seed production and is relatively short, not taking more than 2-3 weeks.

BIOLOGY, ECOLOGY AND PHYSIOLOGY

Oronache spp. are dicotyledonous annual plants (10-60 cm tall, depending upon the species) and recognized by their yellow to straw coloured stems, bearing yellow, white, or blue and snap dragon-like flowers. The leaves are merely triangular scales and both stem and leaves have no chlorophylls. Flowers appear in the axils of leaves and are white and tubular. The fruits are capsules that contain a large number of very small black seeds. *Orobanche* spp. are holo-obligate root parasites, lack chlorophyll and depend completely upon their hosts for carbon, water and nutrients. They are among the most problematic and destructive parasitic species to agriculture in different parts of the world and difficult to control (Rubiales and Fernández-Aparicio, 2012; Habimana *et al.*, 2014; Fernández-Aparicio *et al.*, 2016). These parasites attack a large number of cultivated and wild-grown plant species (Parker and Riches, 1993; Qasem, 2009) and can totally destroy crops under heavy infestation.

Orobanche species depend completely on hosts, while trends in metabolic reduction and dependence of parasitic angiosperms on their hosts for the acquisition of resources is reflected in reduced ability to incorporate inorganic carbon and nitrogen. These parasites lack nitrate reductase and therefore, their ability to utilize ammonium-N depends on the availability of glutamine synthase. *Orobanche* species also depend on hosts for water supply since lack normal root systems, while water flow from host to the parasite is enhanced by the high osmotic pressure attributed to transformation of disaccharides to monosaccharides invertase enzyme. The parasite tends to have high level of mannitol, especially under moisture stress, which may provide a suitable solute that protects the enzyme from denaturation (Parker and Riches, 1993). *Orobanche* spp. accumulate high amounts of mannitol whereas the hexitol is absent from the major host crops (Simier *et al.*, 2001), and thus some inhibitors of the mannose 6-phosphate reductase (M6PR, EC 1.1.1.224), the key enzyme of mannitol synthase, have recently been identified (Robert *et al.*, 1999). In addition, the parasite is rich in starch and can grow slowly for several weeks after the death of its host.

Orobanche negatively affects the host plant but without killing it. It can reduce P, K, RNA, protein levels but increase Ca, Na, Mg and DNA concentration of host plants. In addition, the parasite was found to reduce the total carbohydrate level in host tissues and glutamic acid. Reduction in carbohydrate level may seriously affect the host's ability to absorb water under moisture stress conditions and lead to stomata closure.

With legumes, the most intensive growth of *Orobanche* usually coincides with the flower and pod development of the host. This results in flower dropping and low podding due to increased competition from the parasitic sink.

The initiation of *Orobanche* parasitism and parasite growth and development are affected by different environmental factors and by the development stage of host plants at the time of infestation. These factors include the following:

Temperature: This factor affects parasites at conditioning and germination stages as well as during growth and development. The effect of temperature on germination of different *Orobanche* species indicates differences in their optimum temperature for germination and development, which may be due to differences in their geographical distribution. The lower limit of temperature was found to be 8°C, optimum is about 20°C or slightly higher and the highest is around 30°C. However, some intraspecific differences were reported in that *O. aegyptiaca* and *O. crenata* requires higher temperature at all stages of development. The lower and upper limits of germination temperature may explain some field observations. The time of germination, attachment, and further development of *O. crenata* was not related to the broadbean growth stage rather to soil temperature (Manschadi *et al.*, 2001). In the Mediterranean region, *Orobanche* species are weeds of winter crops. In areas where mid-winter temperatures are below the limits of germination temperature, *Orobanche* germination may be reduced or completely delayed until next spring. Therefore, low temperature decreases the number of germinating parasites and the effect on the host because the position in the host root system is also affected. Reduction in germination at high temperatures may play a role in this context as well. It was found that seeds of *O. ramosa* tolerate temperature up to 50°C for 35 days while air dry, but wet seeds were killed at 40°C within a short time. Exposure to 30°C in 100ppm GA solution appeared to be more damaging than exposure to 35°C in water (Drennan and Mohamed-Ahmed, 1992).

Light: *Orobanche* is affected by light; hence intensity appeared of significant influence on parasite infestation. Reduction in light intensity reduced *O. crenata* and *O. aegyptiaca* infestation to broadbean. However, day length or host flowering appeared with no direct effect on the flowering of *Orobanche* plants. *In vitro*, studies on the effect of light on germination of *Orobanche* species showed indirect intraspecific variation. Certain species seem to be indifferent, whereas others were completely inhibited. Factors other than light also differ and affect *Orobanche* infestation. Low light intensities often coincide with high air humidity. Under humid conditions, the transpiration rate of the host plant is lower, and thus, lower uptake of water and nutrients to the shoots were found. Mohamed-Ahmed and Drennan (1994) reported that a 20 h photoperiod resulted in the earlier emergence of *O. ramosa* and plants of broadbean, lentil or peas were larger than plants grown under a 10 h photoperiod.

Moisture: This is another environmental factor that affects *Orobanche* infestation and prevalence. Soil moisture is necessary for parasite germination and development, and it affects host physiology and growth. In well-drained and textured soils, most of the root parasites find optimal conditions for germination, while in wet soils, germination is poor because of a lack of oxygen for respiration. Germination stimulants may be diluted or washed off under these conditions. However, different workers mentioned *Orobanche* adaptation to dry conditions. Germination took place even at an osmotic potential of -1.29 MPa. Normal angiosperm species are only able to germinate up to an osmotic potential of about -0.4 to 0.5 Mpa. Although certain *Orobanche* species appeared under both rainfed and irrigation conditions, differences exist between different species. *Orobanche cumana* and *O. minor* are mostly associated with rainfed conditions, *O. ramosa* and *O. aegyptiaca* are common in irrigated fields but not exclusively so. *Orobanche crenata* was found more under sufficient than under limiting water supply (Manschadi *et al.*, 2001). In addition, *Orobanche* spp. may be favored by low relative humidity (Parker and Riches, 1993).

Flooding: This factor was sometimes found to reduce *Orobanche* infestation in certain crops. The effect may have various causes, one of which is that germination stimulants get dissolved, diluted, and then inactivated. Another reason may be due to the disappearance of black pigments of *Orobanche* seed coat. Flooding may have an indirect effect by lowering oxygen supply which may cause seed dormancy or seed suffocation or reducing attachment and establishment of the parasite seedlings. Light also affects the growth phase since the parasite can't do photosynthesis and produce oxygen for respiration. Respiration was high in the haustorial region of the host root system. Therefore, if oxygen supply via the host is limited, then flooding will be fatal for *Orobanche,* especially during the hypogeal phase.
Orobanche problem was reported to be less in moist soils and thus in most areas around the Mediterranean where *O. crenata* on broadbean is a real problem under rainfed conditions. Mohamed-Ahmed and Drennan (1994) reported that water logging is an effective cultural practice for *O. ramosa* control. These authors found that the number of emerged *Orobanche* shoots decreased with a longer logging period. Studies on the optimal use of water and its level indicate that for broadbean during pod filling and good yield, soil moisture is essential, but the number of attachments was higher at high moisture levels.

Nutrients: In general, *Orobanche* species prefer poor soils and thus, their parasitism effects could be more severe. Parasitism was found to occur more readily in potting media low in fertility (Jain and Foy, 1992). These conditions were considered important factors associated with severe infestation in the fields (Gharib, 1973).
Animal manure and fertilizers reduced *Orobanche* problems and increased crop yield. Some workers reported that manure application and certain chemical N-fertilizers improved crop yield as a result of their effects on *Orobanche* infestation (Pesch and Pieterse, 1982; Agabawi and Younis, 1985). *In vitro* study it was found that ammonium-N and urea-N had negative effects on germination percentage and growth of *O. crenata* (Pieterse, 1991a; Van Hezewijk and Verkleij, 1996). However, nitrate-N had no such effect while the effect was absent under soil conditions tested(Jain and Foy, 1992). The effect of superphosphate on *O. ramosa* germination in soil with *Carum ajowan* as phosphorus stimulated the germination but hampered the growth of procaulome and thus greatly reduced *Orobanche* attachment to host plants.
The effect of fertilizer may be indirectly due to reduction in the host root system, while other mechanisms may operate as well. There might be a reduction in root exudates of host plants or better defense of the host roots under a suitable supply of minerals. It was suggested that urea-N influences two physiological processes (Pieterse, 1991b). Parasite penetration into broad bean roots was stimulated by the presence of rhizobium, which may be the reason for the low number of *Orobanche* found at high nitrogen levels. The presence of bacteria is necessary for the germination of different *Orobanche* species. *Rhizobium* and mycorrhiza are essential symbionts for Leguminosae and other species. Germination seems to be best in slightly acid soils, which may be due to the interaction with germination stimulants.

The effect of organic (manure and compost) and inorganic fertilizer on *Orobanche* infestation have been long reported (Popova, 1942), and the idea is not new. The primary fertilization effects

are due to nitrogen, especially to ammonium sulfate and urea, while the effect of nitrate was small. However, recent studies on the role of chemical fertilizers are in conflict and showed that extremely high fertilizers (*e.g.* ammonium sulfate) should be applied to achieve some level of control, suggesting that fertilizer affects parasites by dehydrating the germination tube through the plasmolysis process. Westwood and Foy (1999) concluded that ammonium-N was more inhibitory than nitrate form of nitrogen while the elongation of the seedling radicle primarily inhibited by ammonium rather than seed germination and the effect was varied between species. One proposed mechanism of action of nitrogen may be due to direct interaction with parasite metabolism and the alteration in osmotic potential (Ernest, 1986). However, other workers reported that using fertilizer as a control method revealed inconsistent results with respect to parasite growth, with nitrogen treatment often having no discernible effect (Smaling *et al.*, 1991) or even increasing parasite growth (Pieterse and Verkleij, 1991).
Osmotic stress has been suggested as another possible reason for inhibiting *Orobanche* germination by nitrogen (Ernest, 1986). Still, no studies were carried out to correlate this factor to nitrogen inhibition of *Orobanche* seed germination.

Soil type: It has been indicated that *Orobanche* species are in favor of alkaline soils in the Mediterranean region (*e.g.*, *minor* and *ramosa*), but previous contradictory reports indicated that both *O. minor* and *O. ramosa* were in worse conditions on alkaline soils. In addition, no strict link was found between prevalence of *Orobanche* species and high pH, while poor soils are considered most infested. It has been generally observed that *Orobanche* infestation is common in both clay and sandy soils, providing that host plants are found.

Salinity: Few reports showed the effect of salinity on the prevalence of *Orobanche* in certain saline regions. Al-Khateeb (2002) reported that NaCl reduced *O. cernua* germination when seeds were exposed during the conditioning period. NaCl solution (75mM conc.) added to potted soil inoculated with *Orobanche* seeds and grown by tomato completely prevented parasite infestation. Results of this work may support the idea that fertilizers may act against *Orobanche* by affecting or killing the highly sensitive germination tube in a physical rather than any other mechanism.

Insects and pathogens: *Orobanche* species are attacked by *Phytomyza orobanchia* (Diptera, Agromyzidae) (Hassanein *et al.*, 1998). The insect attacks seed capsules (Jafarzadeh and Pourmirza, 1999) and causes a great reduction in seed production of the parasite ranging between 29-94% depending on *Orobanche* species and location (Tsybul'skaya and Skoklyuk, 1978). It also damages the stalk of the parasite. However, the larvae mine in *Orobanche* shoots and capsules. As a consequence, a natural average reduction of *Orobanche* seed production by 30 to almost 80% has been reported from different countries but releases of 500 to 1000 adults/ha resulted up to 96% reduction in *Orobanche* seed production (Klein and Kroschel, 2002). However, this monophagous insect is influenced by different factors including the presence of its natural enemies in the field and the use of insecticides, which may limit its efficiency. In addition, other insects have also been mentioned to attack *Orobanche* plants, including *Psila* species at the larval stage that attack stem of *O. minor* and *Chyliza extenuata*. Other insects parasitize *Orobanche* spp. are polyphagous and often damaging to the crop.
On the other hand, many of the fungi species were reported to attack different *Orobanche* species (Table 1.1.1) including *O. aegyptiaca*, *O. cernua*, *O. crenata* and *O. ramosa* (Parker and Riches, 1993). Only 4 genera proved highly effective and selective to *Orobanche,* these are *Ulocladium* species, *Fusarium* species (*F. Oxysporum* var. *orthoceras*; *F. orobanchas*), and *Alternaria* sp.,

while *Rhizoctonia solani* was found effective against *O. ramosa* in tomato and reduced its infestation by 80%, and *Ulocladium atrum* was effective as a biocontrol agent for *O. crenata* at high humidity.

Table 1.1.1. Fungi species reported to attack different *Orobanche* species

O. aegyptiaca	*Colletotrichum lagenarium*; *Fusarium oxysporum* f.sp. *orthocerus*; *Fusarium arthrosporioides.*
O. alba	*Alternaria tenuis; Alternaria infectoria, Hyphoderma roseum.*
O. cernua	*Fusarium lateritium; Fusarium oxysporum; Fusarium sambucinum; Monilia humicola.*
O. crenata	*Alternaria alternata; Cladosporium herbarum; Cochliobolus spicifer; Fusarium compactum; Fusarium equiseti; Fusarium solani; Fusarium oxysporum* f. sp. *Orthocerus; Fusarium sambucinum; Phomopsis* sp.*; Sclerotinia* sp.*; Ulocladium atrum; Ulocladium botrytis; Ulocladium consortial; Cylindrocarpon* sp.*; Penicillium cyclopodium; Mucor odoratus.*
O. cumana	*Fusarium oxysporum; Fusarium sambucinum.*
O. mutelii	*Cladosporium fasciculatum; Fusarium gibbosum; Fusarium lateritium; Fusarium oxysporum* f. sp. *Orthocerus; Verticillium microsporum; Rhizopus mocrosporus; Rhizopus nigricans.*
O. nana	*Cladosporium fasciculatum.*
O. orientalis	*Aspergillus glaucus; Aspergillus versicolor.*
O. ramosa	*Fusarium lateritium; Gliocladium varians; Rhizoctonia solani.*
O. reticulata	*Conoisporium arundinis; Fusarium lateritium; Fusarium moniliforme.*

Source: Different references

DISTRIBUTION AND HOST RANGE

The Orobanchaceae comprises about 90 genera, including more than 2000 species, most of which belong to the northern hemisphere. Most species occur in the temperate zone over most of the world. They are important in Southern and Eastern Europe, New Zealand, Mediterranean region, the Middle East and North Africa. They have been reported in the USA, Australia and some South America countries. Several hundred thousand hectares are infested in Europe, Middle East, Russia, China, Cuba and California (Parker and Riches, 1993). *Orobanche* originated in Turkey, Italy, Spain, and Morocco. *Orobanche* species are distributed from the Atlantic up to China (*O. cernua*), Cape Verde through North Africa up to Kaukasus and Mozambique (*O. minor*). South of the equator to East Africa and New-Zealand (*O. minor*), West Australia (*O. cernua*) and South Africa (*O. mutelii*).

Orobanche aegyptiaca occurs mainly in Southeastern Europe, Northeastern Africa, and in the Middle East, whereas *O. ramosa* (closely related to *O. aegyptiaca*) is mostly found in the Middle East and introduced into Cuba, Mexico, Australia and finally to California and Texas in the USA (Barney *et al.*, 2005). *Orobanche cernua* and *O. cumana* are spread in the Middle East, infest wide areas in the Mediterranean, Southern and Eastern Europe, and Northern Africa (Torres and Granados, 1991; Munoz *et al.*, 1993). *Orobanche aegyptiaca, O. crenata* and *O. ramosa* are widespread destructive parasites in many countries in the Middle East, including Jordan. However, the natural occurrence of *Orobanche* coincides with the vegetation zone characterized by sclerophyllous woodland with winter rain, natural savanna or grassland areas with mainly 250-500 mm precipitation. Nowadays, *Orobanche* species are reported from more than 58 countries. *Orobanche* spp. are root parasites to many important broad-leaf crops, most damaging to species of Solanaceae, Leguminosae, Asteraceae, Brassicaceae, Lamiaceae, Umbellifereae, Crucifereae and Cucurbitaceae families, and impact directly food destined for human consumption (Ismael and Obeid, 1976; Hodosy, 1981; Sauerborn, 1991; Jain and Foy, 1992; Parker and Riches, 1993; Diaz and Normabuena, 2001). Most species are host-specific, and sometimes restricted to a single species or genus. Others can parasitize a number of unrelated plants, although usually show strong preferences in certain areas. However, there are several species of *Orobanche* each has a wide host range.

Orobanche crenata and *O. ramosa* are most common and destructive species, attack crops of umbellifereae family such as *Daucus carota, Apium graveolens, Petroselinum sativum, Foeniculum vulgare,* and leguminosae including *Vicia faba, Pisum sativum, Lens culinaris, Cicer arietinum* and *Medicago sativa.* In most cases *O. cernua* and *O. ramosa* attack crops of the solanaceae family like *Lycopersicon esculentum, Nicotiana tabacum, Solanum tuberosum* and *Solanum melongena* cause yield loss of more than 75% (Kamel, 2005). The host range of selected *Orobanche* species are summarized in Table 1.1.2.

Orobanche aegyptiaca and *O. ramosa* are more specialized in solanaceous crops, *O. crenata* and *O. minor* in grain and fodder legumes, while *O. cumana* is highly restricted to sunflower. *Orobanche crenata* is the most destructive to legumes and mainly in the Mediterranean region, but it is not important in cucurbits or composite species. *Orobanche aegyptiaca* requires higher temperatures for germination and development than *O. crenata*, and therefore it is more often foundparasitizing tobacco or tomato instead of the legumes.

Table 1.1.2. *Orobanche* spp. and some of their hosts in different countries

Orobanche species	Host species	Reference
Orobanche aegyptiaca	*Anglica archangelica* L. *Anthriscus cerefolium* (L.) Hoffm.) *Apium graveolens var. dulce* (Mill.) DC. *Arachis hypogaea* L. *Armoracia lapathifolia* Gilib. *Aster* spp. *Brassica juncea*. (L.) Czern *Brassica napus* L. *Brassica oleracea* L. *var. botrytis* *Brassica rapa* L. *Capsicum frutescens* L. *Daucus carota* L. *Carthamus tinctorius* L. *Cicer arietinum* L. *Cucumis melo* L.	hohenheim.de/~www380/parasite/control.htm., cited by Qasem, 2006b http://www-aes.tamu.edu/mary/brmrape/brmrape.htm; Abedi *et al.*, 2014; Razavifar *et al.*, 2017; Hammad *et al.*, 1967; http://www-aes.tamu.edu/mary/brmrape/brmrape.htm; http://www-aes.tamu.edu/mary/brmrape/brmrape.htm cited by Qasem, 2006a) Sauerborn, 1991

	Cucumis sativus L.	
	Cucurbita pepo L.	
	Ephedra sp.	
	Foeniculum vulgare Mill.	
	Helianthus annuus L.	
	Lactuca sativa L.	
	Lens culinaris Medik.	
	Lycopersicon esculentum Mill.	
	Matricaria chamomilla L.	
	Nicotiana tabacum L.	
	Ocimum basilicum L.	
	Petroselinum sativum Hoffm.	
	Petroselinum crispum (Mill.) Nym.	
	Salvia sclarea L.	
	Sinapis alba L.	
	Solanum melongena L.	
	Solanum tuberosum L.	
	Trifolium alexandrinum L.	
	Vicia faba L.	
	Vicia sativa L.	
Orobanche alba	Wild plants	Klyueva and Pamukchi, 1979; Baloch *et al.*, 1980
Orobanche alsatica	Wild plants	Klyueva and Pamukchi, 1979
Orobanche arenaria	Wild plants	
Orobanche brassicae	*Brassica oleracea* L. *var. capitata*	
Orobanche caesia	Wild plants	
Orobanche caryophyllacea	Wild plants	
Orobanche cernua	*Helianthus annuus* L.	Chalkov and Shabanov, 1970
	Lycopersicon esculentum Mill.	Munoz *et al.*, 1993
	Nicotiana tabacum L.	Qasem, 2009; Punia, 2014
	Solanum melongena L.	Munoz *et al.*, 1993; Qasem, 2009
Orobanche coerulescens	Wild plants	Klyueva and Pamukchi, 1979

Orobanche crenata	*Anglica archangelica* L. *Anthriscus cerefolium* (L.) Hoffm.) *Apium graveolens* var. *dulce* (Mill.) DC. *Carthamus tinctorius* L. *Cicer arietinum* L. *Daucus carota* L. *Foeniculum vulgare* Mill. *Gazania longiscapa* DC. *Helianthus annuus* L. *Lathyrus odoratus* L. *Lens culinaris* Medik. *Lens esculanta* L. *Pelargonium graveolens* L. *Petroselinum crispum* (Mill.) Nym. *Pisum sativum* L. *Salvia sclarea* L. *Verbena officinalis* L. *Vicia sativa* L. *Vicia faba* L. wild plant	Geipert, 1993 Vorländer, 1988 Linke *et al.*, 1990 Hammad *et al.*, 1967; Tawfik *et al.*, 1976 Nemli and Giray, 1983 Giray and Nemli, 1983 Martelli, 1933a, b http://www-biol.paisley.ac.uk/bioref/Plantae/Orobanche html; http://www-biol.pai; http://www-aes.tamu.edu/mary/brmrape/brmrape.htm; Grandos and Torres 1993; Torres and Granados, 1991 http://www.dijon.inra.fr/malherbo/hyppa/hyppa-a/oracr_ah.htm (Cited by Qasem, 2006a); Qasem, 2009; http://www-biol.paisley.ac.uk/bioref/Plantae/Orobanche.html;http://www-aes.tamu.edu/mary/brmrape/brmrape.htm
Orobanche cumana	*Cannabis sativa* L. *Helianthus annuus* L. *Lycopersicon esculentum* Mill. *Nicotiana tabacum* L. *Solanum melongena* L. *Solanum tuberosum* L.	Trenchev, 1981 Elzein *et al.*, 1999 Mihajlovic, 1986 Lekic, 1970a, b, 1974 Horvath 1981, 1983, 1987; Szökö, 1968 Horvath and Bujaki, 1990 Manjunath and Nagarkatti, 1977 Bronstejn and Kabulov, 1961 Klyueva and Bonduryanskaya, 1986; Klyueva *et al.*, 1972 Klyueva and Pamukchi, 1978, 1979 Kott, 1969 Klyueva and Ursake, 1974 Baloch *et al.*, 1980 Bronstejn and Kabulov, 1961 Petzoldt *et al.*, 1993 Bronstejn and Kabulov, 1961

		Munoz *et al.*, 1993.
Orobanche elatior	Wild plants	Klyueva and Pamukchi, 1979
Orobanche rapum-genistae	*Cytisus scoparius* (L.) Link.	Kroschel own observation, 1995
Orobanche gracilis	Wild plants	Klyueva and Pamukchi, 1979
Orobanche hederae	*Hedera helix* L.	Martelli, 1933b Klein own observation, 1995
Orobanche lutea	Wild plants	Klyueva and Pamukchi, 1979
Orobanche major	Wild plants	Klyueva and Pamukchi, 1979 Horvath and Bujaki, 1990
Orobanche minor	*Apium graveolens* L. *Arachis hypogaea* L. *Carthamis tinctorius* L. *Daucus carota* Hoffm. Arcang *Guizotia abyssinica* L.F. Cass. Geraniaceae (including Ornamental *Pelargonium* spp.) *Helianthus annuus* L. *Hypochaeris* spp., *Lactuca sativa* L. *Lens culinaris* Medik. *Lotus comiculatus* L. Malvaceae family members, *Medicago sativa* L. *Nicotiana tabacum* L. *Pisum sativum* L. Rosaceae spp. *Trifolium pratense* L. *Trifolium repens* L. *Trifolium subterranean* L. *Vicia faba* L. *Vicia* spp.	Miller *et al.*, 1997; http://www- aes.tamu.edu/mary/brmrape/brmrape.htm; Dhanapal, 1996 Dominguez *et al.*, 1993. http://www.dijon.inra.fr/malherbo/hyppa/h yppa-a/oracr_ah.htm, Hamid *et al.*, 2000. Miller *et al.*, 1997;http://www- biol.paisley.ac.uk/bioref/Plantae/Orobanch e.htmlhttp://www-biol.pai; http://www- biol.paisley.ac.uk/bioref/Plantae/Orobanch e.html;http://www- es.tamu.edu/mary/brmrape/brmrape.html; Dhanapal, 1996; Watzl, 1936; Klyueva and Pamukchi, 1979
Orobanche mutelii	*Arachis hypogaea* L. *Armoracia lapathifolia* Gilib. *Brassica oleracea* L. *var. capitata* *Helianthus annuus* L. *Lactuca sativa* L. *Lycopersicon esculentum* Mill. *Nicotiana tabacum* L. *Solanum melongena* L. *Solanum tuberosum* L.	Chalkov and Shabanov 1970; Klyueva and Bonduryanskaya, 1986; Klyueva *et al.*, 1972

Orobanche pallidiflora	Wild plants	Klyueva and Pamukchi, 1979
Orobanche purpurea	Wild plants	
Orobanche ramosa	*Amaranthus retroflexus* L.	University of California, 1965;
	Apium graveolens L.	http://www.dijon.inra.fr/malherbo/hyppa/h
	Artemisia campestris L.	yppa-a/oracr_ah.htm;
	Beta vulgaris L.	http://www-
	Brassica napus L.	aes.tamu.edu/mary/brmrape/brmrape.htm;
	Brassica oleracea L. *var. capitata*	University of California, 1965; Brault *et al.*
	Cannabis sativa L.	2007; Vouzounis, 2005; Benharrat *et al.*
	Capsella bursa-pastoris (L.)	2005
	Medicus	University of California, 1965
	Capsicum annuum var. grossum	
	(Willd.) Sendtn	http://www-
	Citrullus vulgaris Schrad.	biol.paisley.ac.uk/bioref/Plantae/Orobanch
	Coleus scutellarioides Benth	e.html; http://www-biol.pai;http://www-
	Cucumis melo L.	aes.tamu.edu/mary/brmrape/brmrape. html
	Cucumis sativus L.	hohenheim.de/~www380/parasite/control.h
	Cucurbita pepo L.	tml
	Cucurbitaceae family	University of California, 1965
	members	
	Engelmannia peristenia (Raf.)	
	Goodman	
	Geranium maculatum L.	
	Helianthus annuus L	
	Lactuca sativa L.	
	Leguminosae and several	Chalkov and Shabanov, 1970
	other families.	Trenchev, 1981
	Lupinus subcarnosus Hook	Kott, 1969
	Lycopersicon	Elzein *et al.*, 1999
	esculentum Mill.	Mihajlovic, 1986,
	Medicago polymorpha L.	Lekic, 1970a, b, 1974
	Members of Cruciferae,	Horvath, 1981, 1983, 1987; Szökö, 1968
	Nicotiana tabacum L.	Moiseeva and Mamraliev, 1969
	Phaseolus vulgaris L.	Horvath and Bujaki, 1990
	Solanum melongena L.	Al-Khesraji *et al.*, 1987
	Solanum nigrum L.	Bronstejn and Kabulov, 1961
	Solanum spp.	Klyueva and Bonduryanskaya, 1986
	Solanum tuberosum L.	Klyueva *et al.*, 1972
		Klyueva and Pamukchi, 1978, 1979
	Tecurium sp.	Klyueva and Ursake, 1974
	Vicia faba L.	Geipert, 1993
	weeds	Bronstejn and Kabulov, 1961

	wild plants	
	Xanthium spinosum L.	Kott, 1969
	Zea mays L.	Baloch *et al.*, 1980
		Qasem, 2009
		Dhanapal, 1996; Barney *et al.*, 2005
Orobanche sp.	*Brassica napus* L.	Spencer, 1973
	Brassica oleracea L. *var. capitata*	Spencer, 1961
	Cannabis sativa L.	Klein own observation, 1995
	Cicer arietinum L.	Kaltenbach, 1864
	Citrullus vulgaris Schrad.	Spencer, 1973
	Cucumis melo L.	
	Cucumis sativus L.	
	Daucus carota L.	
	Genista sp.	
	Hedera helix L.	
	Helianthus annuus L.	
	Lens culinaris Medik.	
	Lycopersicon esculentum Mill.	
	Nicotiana tabacum L.	
	Pisum sativum L.	
	Solanum melongena L.	
	Solanum tuberosum L.	
	Vicia faba L.	
	Wild plant	

ECONOMIC IMPORTANCE

Although *Orobanche* is spread all over the world, losses occur mainly in South and East Europe, North Africa and West Asia. In the Mediterranean area and West Asia, about 16 million ha of arable land has been reported as threatened by *Orobanche*, which accounts for 1.2% of the world's arable land. Parasitism leads to dependency on the number of parasite attachments per host plant and the time of infestation that leads to more or less intensive yield reduction. Damage by the parasite is dependent on the infestation level in the field, but in most cases it is destructive. Most of the attached parasites hide below the soil usually, only about 1-30% of the attached plants emerge while heavy infestation occurs below the soil at which most of the parasitism occurs and the parasite could be easily observed on the root system at different developmental stages. *Orobanche* spp. attack mainly Asteraceae, Brassicaceae, Apiaceae, Leguminosae, or Solanaceae (Fernández-Aparicio1 *et al.*, 2016). Yield loss was estimated at hundreds of millions of dollars annually, which affect the life of 100 million farmers. Host plants are severely affected, poorly growing and usually fail to flower. *Orobanche* exhausts host plants since it deprives host plants of water, minerals, nutrients and carbohydrates. Because energy deficiency in the host roots causes drought stress and wilting of the host and visible damage, including stunted shoot growth, leaf chlorosis, reduced photosynthesis that results in the flow of carbohydrates to the parasite, reducing host growth and yield quantity and quality. Most of the damage occurs below the soil and before parasite emergence, during which more than 80% of the *Orobanche* dry matter is deposited before flowering. Damage to the host plants is greatest where water and nutrients are limited, which occurs most in regions where parasites are present. Under heavy infestation, crop yields are greatly reduced or completely lost, especially in marginal areas or in dry regions where a few *Orobanche* species are destructive (Silverside, 2002a,b; Barney *et al.,* 2005). However, estimated yield losses range from 5% to 100% depending on host susceptibility, level of infestation, time and duration of infestation and environmental conditions (Stewart and Press 1990) while damage to crop plants from *Orobanche* can range from zero to complete crop failure (Dhanapal *et al.* 1996). In the Middle East, annual losses in food crops caused by *Orobanche* spp. were estimated at billion dollars (Aly, 2007).

Branched broomrape (*Orobanche ramosa*) is the most destructive species. In heavily infested areas, this parasite can cause total crop loss (Dhanapal, 1996; Barney *et al*., 2005; Qasem, 2019; 2020a). It is one of eleven *Orobanche* species reported in Jordan (Musselman, 1998) and of eight destructive species of agricultural land found in the country that parasitize many field crops and certain species of fruit and forest trees (Qasem, 2009). It is the most prevalent and devastating species of vegetables in both irrigated and rainfed agriculture. In addition, it attacks a wide range of wild and cultivated plant species of different plant families (Qasem, 2009). It is destructive to tomatoes (*Lycopersicon esculentum,* Mill), the main vegetable crop in many countries. Mauromicale *et al.* (2008) reported that *Orobanche* plants do not only decrease the above-ground mass of tomatoes, but also influence and decreasing chlorophyll mass of the leaf and photosynthesis activity. *Orobanche cernua* is another species that severely attacks tomato, tobacco, and sunflower, with a devastating effect on the crops. Yield loss in tobacco ranges from 30 - 70% in India (Dhanapal, 1996). It is exhaustive to beans (*Phaseolus* sp.) in Italy causes great economic losses and crops, specifically in southern Europe. *Orobanche* causes considerable yield losses in tobacco, clover, tomatoes, sunflowers (*Helianthus annuus* L.), and broad beans in The USA (Ristau, 2001). Heavy crop infestations in beans have been reported from Egypt. *Orobanche cernua* causes a "near wilt" appearance in tobacco and affects yield and quality and heavy infestations destroy 25% of the crop in India (Mariam and Suwanketnikom, 2004). Yield losses due to *O. crenata* may reach as high as 75-100% in certain parts of the Sudan and Ethiopia (ICARDA, 2006), while seed-yield losses caused by *O. foetida* infestation in broadbean reached 95% (Bouraoui *et al.*, 2016).

Orobanche species may also serve as hosts or carriers for different agricultural pests including viruses, bacteria and polyphygous insects. As an example, *O. ramosa* is a new host for the epiphytic development of *Pseudomonas syringae* pv. Tomato (Nevena and Senka, 1997) and has been reported as a novel host with the ability to facilitate the replication and processing of viroids such as PSTVd (Vachev *et al.*, 2010).

On the other hand, *Orobanche* species may have some positive values (Qasem, 2020b). *Orobanche crenata* is regarded as a tasty vegetable plant of high demand by people in certain regions in Italy and can be served in several recipes and offered in restaurants (Rubiales, 1999). *Orobanche* is also grazed by cattle, and parasitic plants including *Orobanche* have been reported to increase milk fat (Kovačević, 1976) and thus cattle serve as an agent responsible on its spread by different means. A few species in Orobanchaceae are used in food and folk medicine (Qasem, 2020b; Shi and Li, 2020).

Great broomrape (*O. major*) was prescribed in Europe for bladder and kidney stones and provoke lustry urine. The juice of this parasite was regarded as the only treatment for old green wounds-the bleeding ulcers and sores-malignant and scabby ulcers, those that are hollow also" and "for fretting sores". Flower spikes decoction was also used for skin cleansing and as a cure for freckles, black or blue spots or push thereof.

CONTROL

It looks that the autotrophic period of *Orobanche* and establishment on host plant is the best stage for control of these parasites (Salle *et al.*, 1984; Nickrent and Musseilman, 2004). Different workers reviewed literature and proposed strategies for *Orobanche* spp. control (Parker, 1991; Parker and Riches 1993; Rubiales and Fernández-Aparicio, 2012; Punia, 2014; Fernández-Aparicio *et al.*, 2016; Samejima and Sugimoto, 2018).

Control methods are many and varied but very limited in effectiveness and consistency. These include quarantine or prevention methods that prevent parasite introduction; agricultural such as planting date, crop rotation (Al-Menoufi, 1991) with trap and catch crops, nitrogenous (Pieterse, 1991; Jain and Foy, 1992) and mineral fertilizers, managed fallow; physical such as the use of soil plastic mulch, soil solarization (Sauerborn *et al.*,1989) with or without animal manure fermentation, flooding, organic material and burning, breeding for resistant or partial resistant cultivars (Foy *et al.*, 1988; Cubero and Hernandez, 1991; Qasem and Kasrawi, 1995); mechanical methods including hand weeding, weeding after harvest and deep ploughing; biological control method using insects and fungi (Zahran, 1973; Garcia-Torres and Lopez-Granados, 1991a) such as *Fusarium oxysporium* (Cubero *et al.* 1999; Lu *et al.* 1999; Thomas *et al.* 1999a;b); chemical methods using soil fumigation (Foy *et al.*,1989) or disinfectants, herbicides, natural or synthetic germination stimulants (Saghir, 1986; Qasem, 2006b) and herbicides. Studies on the effect of systemic herbicides against different *Orobanche* species are very limited, and few have reported promising results with soil-applied herbicides (Foy *et al.,* 1989; Parker and Riches, 1993). However, the use of herbicides has not generally proven successful in the past (Kasasian and Parker, 1971; Kasasian, 1973; Jacobsohn and Kelman, 1980; Foy *et al.,* 1988), although some satisfactory chemicals have been reported as effective against *Orobanche ramosa* L. in celery (Americanos, 1991), tobacco (Lolas, 1986) and potato (Haidar *et al.*, 2005a, b).

In the last two decades, sulfonylurea herbicides were introduced in the struggle against these parasites in tomatoes and results in selective control of *Orobanche aegyptiaca* Pers. by chlorsulfuron and triasulfuron (Ghannam *et al.*, 2012) and *O. ramosa* by chlorsulfuron (Qasem, 1998) have been reported. However, persistence, toxicity and low selectivity of this group and imidazolinones herbicides associated with the high cost and increase in public concern on synthetic agricultural pesticides and their effects on human health and the environment necessitate further attention and search for alternatives in parasites management in different parts of the world.

Ultimately, the development of effective and low-cost control measures for *Orobanche* remains the main goal for weed control specialists, plant pathologists, agronomists, and biotechnologists. Genetic engineering has been recently proposed and can be explored for *Orobanche* management. However, complete control methods are not yet available, although many are usually employed by farmers worldwide, including physical, agronomical, biological, biotechnological and chemical (Qasem, 2006a; Rubiales and Fernández-Aparicio, 2012; Habimana *et al.,* 2014; Disciglio *et al.*, 2016; Fernández-Aparicio *et al.*, 2016). None of these, however, is effective in preventing or completely controlling *Orobanche* spp. (Disciglio *et al.*, 2016; Fernández-Aparicio *et al.*, 2016). Until now, developments in control strategies are not advancing beyond agronomic practices, resistant varieties and possible chemical control by herbicides (Rupialis *et al*, 2018). Therefore, integrated methods through varying agronomic measures are always recommended for a somehow acceptable level of parasite control and to avoid total yield loss under heavy infestation (Parker and Riches, 1993; Qasem, 2019, 2020a). Taking into consideration the cost, persistence, low selectivity, and negative environmental impacts of some recently reported effective sulfonylurea and imidazolinones herbicides against certain parasite species, alternative eco-friendly methods are necessary (Qasem, 2019). These may be recently thought through the use of safe, natural products, plant materials, and their byproducts, allelopathy, resistant varieties, trap and catch species in rotation with host crops, botanical herbicides, growth promoters, and fertilizers, mechanical and physical methods including plastic mulch and soil solarization and even eating broomrape might be considered as part of the integrated control package (Rubiales, 1999). All are strongly recommended for parasites and other weeds control in general.

Prevention Methods

Seeds of *Orobanche* are spread by irrigation water and can transfer with contaminated crop seeds, soil, organic materials, transplants, agriculture machinery, contaminated containers, wind, and workers (Kleifeld, 1998; Karnkowski, 1999).

The best control approach is prevention, i.e., sowing clean seed. *Orobanche* seeds are among the tiniest in existence, so they are easily removed from large-seeded crops. Cleaning crop seeds to remove the minute seeds of *Orobanche* will not normally be feasible, but absolutely necessary and risks could be greatly reduced by very thorough washing. Any bag or container that is known or thought to have been contaminated with *Orobanche* seeds should not be used to transport crop seeds from infested to uninfested areas. One of the most important control methods is to prevent the distribution of parasite seeds from one field to another and from infested to uninfested areas (Kebede and Ayana, 2018). Different mechanisms are responsible for the dispersal of the seeds. National and international trade of crop seeds contribute to seed dispersal over long distances.

Other methods of prevention of seed dispersal are to prevent animals from feeding on the parasitic weeds or controlling their movement, and to clean shoes and tools when leaving parasitic weeds infested fields. *Orobanche ramosa* L. subsp. *mutelii* (F.W. Shultz) Cout. was reported to transmit by sheep serve as vectors for long-distance dispersal by seed retained in wool or pass through the gastrointestinal tract (Ginman *et al.*, 2015). Moreover, it is important to recognize and to eradicate parasitic weeds in new infested fields to prevent any build-up of the parasitic weed population. Hand pulling and burning of weeded parasite shoots is important to prevent the reproduction of thousands of seeds (Miller *et al.*, 1997) and control of weeds in general during fallow years and before seeding looks also necessary to prevent parasitic weed spread on wild or weed species as well as weed infestation.

Cultural and Mechanical Methods

These include the following:

Land Preparation

There are periodic reports of the advantages of deep ploughing and zero or minimum tillage. Deep ploughing was suggested by Labrada and Perez (1988) for *O. ramosa* in tomato in Cuba and by several authors for *O. cernua* in tobacco (Okazova *et al.*, 1980; Khot *et al.*, 1987; Krishnamurthy, 1987), but the problems are the cost and practicality of the initial treatment and the tillage practices in subsequent years do not seem to have been adequately considered. Zero and minimum tillage increased broadbean infestation by *O. crenata* (Kukula and Masri, 1984), while earlier reports mentioned that ploughing at 40-50 cm deep reduced *Orobanche* by 80-90% in three year-period (Cited by Kasasian, 1971).

Planting or Sowing Date

The importance of planting date on the intensity of *O. crenata* infestation has been emphasized by several authors. Mesa-Garcia and Garcia-Torres (1982) in Spain, Raaimakers *et al.* (1988) in Egypt and Van Hezewijk *et al.* (1987) in Syria have all confirmed how early planted broadbean and lentil (*Lens culinaris* Medik.) are more severely infested, apparently because low soil temperatures reduce *O. crenata* germination on the later planting dates. However, delayed planting greatly reduces potential crop yield. According to Munoz *et al.* (1993) the duration of *O. cernua* emergence period increased as the sowing date was delayed. Since sunflower yield was higher at earlier sowing dates, early sowing of this crop may be recommended in *Orobanche*-infested areas under mild Mediterranean climate. Germination of *O. crenata* tends to be zero below 8°C, and further development is greatly reduced at low temperatures. Delaying the planting date affects *Orobanche* more than its hosts; the delay should be only two weeks from the optimal date for sowing in an uninfested field (Habimana *et al.*, 2014). This delay will result in partial avoidance of infestation, as environmental factors affect the germination and development of the parasite. It was found that for effective control of *O. crenata*, broadbean cv. Giza 402 should be sown around mid. November in Egypt without glyphosate application and with zero tillage (Nassib *et al.*, 1984). Early sowing of peas effectively reduces *Orobanche* incidence (Pirani *et al.*, 1993) and early sowing of sunflower may be recommended in *O. cumana* infested areas under Mediterranean climates (Castejon-Munoz *et al.*, 1993). However, change of the sowing date seems not to be very promising under certain conditions, but it is still recommended in different regions. *Orobanche cumana* on sunflower appeared earlier at late than early planting dates. *Orobanche* shoots number per plant was lower when sunflower planting was late (Aydin and Mutlu, 1996). In another study, parasitism of *O. crenata* was decreased with late sowing dates, while better results were obtained with resistant varieties and late sowing (Sillero *et al.*, 1999). Perez-de-Luque *et al.*, (2001) came also to the same conclusion.

Planting Depth

The position of *Orobanche* seeds in the soil plays a major role in determining parasitism. Mohamed-Ahmed and Drennan (1994) found that *O. ramosa* significantly failed to emerge when crop plants were grown from seeds planted at 11-cm or deeper than seeds planted at 3-4 cm depth. Eltayeb *et al.* (2000) reported that *O. ramosa* seeds placed 10 cm deep resulted in heavy infestation compared to those placed at 15 cm; thus, tomato seedlings at 15 cm reduced parasite infestation.

Hand Weeding

Weeding parasitic weeds is an option available to farmers worldwide with different costs and less so in populated poor societies. The use of this practice should be the first priority of extension effort. Hand-pulling of *Orobanche cernua* in tobacco has been recommended in India and is claimed to result in almost complete control after 3 years (Krishnamurthy and Rao, 1976). It may be less satisfactory in other crops because of difficulty of access, as under a dense canopy of tomato.

Clean cultivation followed by hand weeding or hoeing *Orobanche* plants before producing and disseminating their seeds were commonly recommended for control. This was especially important with new infestations. The attachment with the host roots is not very firm, and the whole parasite breaks away readily. It is necessary to remove broomrape plants systematically for several years until the dormant seeds have all developed, taking into consideration that survival of only a very few plants is sufficient to reinfest a large area (http://www.wssa.net/subpages/weed/larrymitich/Orobanche.html).

However, weeding parasitic weeds is not very promising in highly infested areas, but it is an alternative method available to all farmers and should be first implemented, though its exact function and limitations should be clearly understood. Certain workers insisted on the application of this method of control even in the field with dense infestation. In the short term, economic analysis of weeding is likely to show no benefit because of little or no yield increase.

Weeding is time-consuming, labor input is too high in certain regions, and its success is only in a long term and can cause injuries to the root system allowing the invasion of certain soil-borne pathogens. Since no single method is reliable enough, hand pulling of remaining *Orobanche* shoots as an accompanying method is most important to avoid re-infestation. In combination with other methods, it can reduce the seed bank effectively and therefore weeding is an economical method on a long-term basis.

Knowing more about the reproduction of *Orobanche* by innumerable seeds will lead to a better acceptance of hand weeding as an important tool of control, especially in fields with starting infestation. Using a manual spear (iron blade stick) in tobacco to cut vegetative shoots of *O. cernua* at ground level or 2-3 cm deep in the soil and repeated 3-4 times a season at 4-5 days intervals was effective in eradicating this parasite within 5-6 years (Krishnamurthy, 1992).

Crop Rotation with Trap/Catch Crops

Trap Species

The fast dispersal of *Orobanche* into new fields is related to crop rotation. Monocropping or frequent cropping with host crops speeds up the process. Cultural methods usually involve the inclusion of trap or catch crops that normally alternate with crop species in rotation. Crop rotation with non-host crops, especially with trap species, is of great advantage (Eltayeb *et al.*, 2000).

Crop rotation is already a current farmers' practice, and susceptible crops are not cultivated year by year on the same fields. However, different workers reported successful control of different *Orobanche* species following this method. Schnell *et al.* (1994) reported legume crops to cause a greater reduction in *O. crenata* seed banks. Resistance to *O. crenata* has been found both in cultivated and wild-grown legumes (Rubiales *et al.*, 1998; 1999; Sillero *et al.*, 2001a,b). More recently, resistance to *O. crenata* has been identified in different species of *Lathyrus* and has suggested that these could be regarded as trap crops to be used under field conditions (Sillero *et al.*, 2001a).

The ability of certain plant species to stimulate *Orobanche* seed germination encouraged research on the possible inclusion of catch and/or trap species to reduce parasite infestation. Trap species is a false host that stimulates the parasite seed germination but does not support parasitism. Thus, it works in a suicidal way that leads to exhaustion of parasite seed banks when integrated with other methods of parasite management. It is employed to deplete the seed reserve; this is referred to as a seed germination at which either traps species or stimulatory natural chemicals (Yoneyama *et al.*, 1998; Evidente *et al.,* 2010; Daniel *et al.*, 2011) may be incorporated in parasite management. When broomrape seed occurs close to the roots of such species, 13 to 15% of the seeds germinate but do not develop haustorial connections. Suicidal germination of the parasitic seeds triggered by growing trap crops reduced weed population, and growth of the host plants was encouraged due to the green manuring effect of trap crops. Therefore, including a trap crop in the rotation may reduce the problem. Sunhemp (*Crotalaria juncea* L.) and green gram (*Vigna radiata* L.) are promising trap crops in a cropping system containing bidi tobacco in areas where tobacco is grown in a long growing season. Peppers grown in rotation with tobacco tend to reduce broomrape seed infestations in the soil (https://wssa.net/wp-content/themes/WSSA/WorldOfWeeds/orobanche.html). Pepper (*C. annuum*) roots stimulated germination of 22-26% of nodding *Orobanche* seeds but without the formation of parasitic attachments. Four-fold increase in the number of nodding *Orobanche* was observed on tomato roots interplanted with pepper compared with the number of parasites on tomato roots when planted without pepper. Indicating that some of the parasite seeds enhanced by pepper were attached to tomato roots since pepper works as a trap species. In a study carried out by Hershenhorn *et al.* (1996) few attached and developed *O. aegyptiaca* shoots were found on sweet pepper. *Vicia villosa* subsp. *dasycarpa* in rotation with lentil, chickpea and broabean kept *O. crenata* infestation at a low level (Schnell *et al.*, 1996). Other studies reported that growing flax in two successive winter seasons or one summer cropping with *Phaseolus aureus* Roxb. (*Vigna radiata*) reduced early infestation of *O. aegyptiaca* and significantly increased tomato growth and production (Kleifeld *et al.*, 1994). Other species reported including coat buttons (*Tridax procumbens* L.) and hairy beggar-ticks (*Bidens pilosa* L.) (http://www.wssa.net/subpages/weed/larrymitich/Orobanche.html, cited by Qasem, 2006a).

Some of the species proved highly successful in rotation included sesame, brown Indian hemp, common flax and black-eyed pea (*Vigna unguiculata* (L.) Walp.) that decreased broomrape biomass by 86, 85.3, 75.2, and 74.4%, respectively and resulted an increase in the tomato yield. Sesame, brown Indian hemp, Egyptian clover and mungbean increased total biomass of tomato by 71.4, 67.5, 65.5, and 62.5 %, respectively (Babaei *et al.*, 2010). Flax was the most effective and reduced *O. crenata* shoot number and dry weight by 52% and 71% in the first year and by 55% and 26% for the second year in lentil fields, respectively. *In vitro,* flax seed exudates induced germination of *O. crenata* and *O. ramosa,* and similar results were obtained on *O. aegyptiaca* with flax root exudates, GR-7 and GR-24 (Matthews *et al.*, 1991). However, flax as a trap crop for eliminating *Orobanche* infection in broadbean fields was mentioned to has a limited potential (Khalaf, 1992). Flax and sugarcane significantly reduced *O. crenata* in rotation with broadbean (Abou-Salama, 1995). Broccoli also reduced shoot number of the same parasitie species by 48% and 39% in two years (Aksoy *et al.*, 2016). Different species have been reported to show strong ability' to stimulate seed germination of different *Orobanche* species by more than 90%, and extracts of hundreds plant species were tested for possible stimulating or inhibiting seed germination of different *Orobanche* spp. and many may be considered as trap, cover, catch species or as a source of natural germination stimulants for these parasites (Babaei *et al.*, 2010; Ma *et al.*, 2012; Qasem, 2019). Trap crops reduce parasite seed bank in the soil and cause a decrease in infestation. They stimulate only parasite seed germination, without becoming parasitized and do not require any additional labor when included in crop rotation. Heavily infested fields should be planted with trap species for 2 or 3 years to exhaust the parasite seed bank.

Different cultivated or wild species, including plants producing ethylene or strigoles with a great potential to stimulate seed germination of certain parasitic weeds, including different species of *Orobanche,* without being themselves attacked, have been reported and trap species for different *Orobanche* species are shown in Table 1.1.3.

In a study carried out by Chabrolin (1935), 65 nonhost plant species able to stimulate *Orobanche* seed germination were identified. Other *Orobanche* stimulating species reported were sorghum (*Sorghum vulgare* Pers.), barley (*Hordeum vulgare* L.), and vetch (*Vicia dasycarpa* spp. *villosa*) (Kasasian, 1973; Linke *et al.*, 1991) for *O. crenata*; bean, sorghum, maize, and cucumber for *O. ramosa* (Labrada and Perez, 1988) and sorghum, cowpea, chili, hemp, mung bean, flax, lucerne, soybean and chickpea for *O. cernua* (Krishnamurthy and Rao, 1976; Krishnamurthy *et al.*, 1977). However, flax was the most potent trap crop. Linke *et al.*, (1991) reported a significant reduction in *O. crenata* infestation after three years of growing *Vicia dasycarpa* spp. *villosa*. Forage legumes have been recommended as trap crops to decrease the broomrape's seed bank in the soil (Linke *et al.*, 1993; Saxena *et al.*, 1994) and to replace fallow in areas where cultivation of other food legumes is not possible. Sillero *et al.*, (2001a) reported the possible use of *Lathyrus choranthus* as a trap crop to reduce broomrape seed bank in the soil as permitted very low emergence of broomrape shoots. Still, they allowed a relatively high establishment of the parasite with low susceptibility to *Orobanche.*

The non-suitable host crops of *O. cernua,* including chili, sorghum, cowpea, *Phaseolus aconitifolius* and *Hibiscus sabdariffa* stimulated germination of the parasite at a high level, while sesame stimulated germination but without allowing further growth and development (Krishanmurthy and Chandwani, 1975). Different studies showed that root diffusate of *Linum usitatissimum*, *C. annuum*, *Sesamum indicum* L. and *Trifolium alexandrinum* enhanced *Orobanche* seed germination, but the germinated seeds normally failed to attach to the root systems of these crops (Abu-Shakra *et al.*, 1970; Krishnamurthy *et al.*, 1977; Musselman, 1980; Al-Menoufi, 1991). Strong induced germination of *O. cernua* seeds was obtained using trap crops, especially *Vigna radiata* and *Crotalaria juncea* (Dhanapal *et al.*, 1998). In addition, good inhibition and resistance level to *Orobanche* infestation to different weed species were detected. In a study conducted by Boulet *et al.,* (2001), in which 30 weed species were tested for susceptibility to *O. ramosa*; they found that *Avena sativa* ssp. *sterilis*, *Ammi majus*, *Solanum nigrum*, and *Anagallis arvensis* possessed different levels of resistance lead to the death of the parasite. These authors suggested using these species as trap plants. Out of 21 tested crops by Rodriguez-Ojeda *et al.* (2001) seven were found stimulating germination of *O. cumana* seeds; these were (from the strongest to the weakest) corn, sorghum, millet, cotton, rice, eggplant and cauliflower.

However, in a short-term cropping system, trap species may increase crop yield, as found by Zemrag and Bajja (2001), who reported that fenugreek and coriander decreased the number of attached parasites per host plant and disturbed their development. They showed that these trap crops increased broad bean yield by 18.37 and 14.82%, respectively.

Maize and snap beans are used as traps for *O. ramosa* and complement each other under inter-cropping. Both depleted *O. ramosa* and *O. cernua* seed bank by 72.5% per season (Abebe *et al.*, 2005) and significantly increased tomato yield in the 3rd season. Dongola (2006) reported that crop rotation of onion-onion-onion-tomato and/or onion-alfalfa-alfalfa-tomato decreased *O. ramosa* infestation by 90-95% and the yield of tomato was increased by 60%. Crop sequences in the two locations, in addition to the increase of tomato yield, decreased the *Orobanche* incidence, branching and seed production. Qasem (2019) reported that *O. ramosa* shoot number was the highest on tomato followed *Anethum graveolens* L., *C. vulgaris*, *Cucumis melo* var. *flexosus* L., *Pimpinella anisum* L., *S. indicum* L., *Solanum elaeagnifolium* Cav., *Sorghum vulgare* Pers., *Spinacia oleracea* L., and *T. alexandrinum* and the lowest after *Brassica oleracea* L. *var. italica* Plenck, *Brassica rapa* L. *var. rapa*, *C. annuum, Capsicum frutescens* L., *Cicer arietinum* L., *Citrullus colocynthis* (L.) Schrad., *Cucurbita maxima* Duch., *Cuminum cyminum* L., *Hordeum vulgare* L., *Linum usitatissimum* L., *Spinacia oleracea* L. cv. Epinard greant and *Vigna sinensis* (L.) Savi. Parasite dry weight per shoot was the lowest on tomato grown after *C. arietinum, C. frutescens, Cucumis melo* L., *Hibiscus sabdariffa* L., *P. anisum* and *T. alexandrinum* but the highest after *Cichorium endivia* L. *var. crispum* Lam., *Peganum harmala* L., *S. oleracea* cv. Epinard greant, and *Zea mays* L. Tomato shoot dry weight increased by 126% over parasite-free control following *Ecballium elaterium* (L.) A. Rich. and parasite infestation reduced by 56% of the *Orobanche*-infested control. High tomato growth and best parasite control (73% reduction) were obtained after *V. sinensis*. *Hibiscus sabdariffa, Hordeum vulgare,* and *S. vulgare* reduced both *Orobanche* infestation and tomato growth.

Stimulants however, are mainly natural chemicals released with root exudates of host and nonhost trap species into the surroundings, received by parasite seeds and stimulate their germination. Some of these chemicals are strigol isolated from maize root exudates, sorgolactone, isolated from sorghum (Hauck *et al.*, 1992), alectrol produced by cowpea and orobanchol by red clover roots (Xie *et al.*, 2008). Some of these chemicals, such as coumarin-type compounds and other

Table 1.1.3. Trap crops for *Orobanche* spp.

Scientific name	Common name	References
Orobanche ramosa		
Allium cepa L.	Onion	Dongola, 2006
Allium sativum L.	Garlic	Abebe *et al.*, 2005; Qasem and Foy, 2007; Qasem, 2006a
Arabidopsis thaliana (L.) Heynh	Thale cress	Goldwasser *et al.*, 2000
Brassica juncea (L.) Czern	Brown mustard	Cited by Kasasian, 1973
Brassica napus L.	Oil seed rape	
Brassica rapa L.	Turnip	Al-Menoufi and Adam, 1996
Capsicum annuum L.	Pepper	Cited by Kasasian, 1971; Sand, 1983
Carum ajowan Benth. et Hock	Bioshops weed	Qasem and Foy, 2007
Carum carvi L.	Carawy	
Coriandrum sativum L.	Coriander	Al-Menoufi and Adam, 1996
Cucumis sativus L.	Cucumber	Labrada and Perez, 1988
Glycine max (L.) Merr.	Soyabean	Qasem, 2006b
Lablab purpureus (L.) Sweet	Hyacinth bean	
Lathyrus ochrus (L.) DC	Winged vetchling	
Linum usitatissimum L.	Linseed	Eplee, 1984; Khalaf, 1992
Lupinus termis Forssk.	Lupine	Al-Menoufi and Adam, 1996
Medicago sativa L.	Lucerne	Ballard, 1979; Ghotbi *et al.*, 2012
Pennisetum sp.	Fountain	Cited by Kasasian, 1971
Phaseolus aureus Roxb.	Green gram	Kleifeld, 1996
Phaseolus vulgaris L.	French bean	Labrada and Perez, 1988; Abebe *et al.*, 2005

Pisum sativum L.	Pea	http://www.uni-hohenheim.de/~www380/parasite/oro_path.htm
Ricinus communis L.	Castor bean	Cited by Kasasian, 1971;
Sesamum indicum L.	Sesame	Qasem, 2006a
Setaria sp.	Millets	Cited by Kasasian, 1971
Sorghum bicolor (L.) Moench	Sorghum	Labrada and Perez, 1988
Trifolium sp.	Clover	Cited by Kasasian, 1971
Trigonella foenum- graecum L.	Fenugreek	Al-Menoufi and Adam, 1996
Vigna radiata (L.) Wilczek	Green gram	http://www.uni-
Vigna unguiculata spp. (L.) *unguiculata*	Cowpea	hohenheim.de/~www380/parasite/oro_path.htm
Zea mays	Maize	Cited by Kasasian, 1971; Labrada and Perez, 1988, Abebe *et al.*, 2005

Orobanche aegyptiaca

Allium sativum	Garlic	Acharya, 2012
Arabidopsis thaliana	Thale cress	Goldwasser *et al.*, 2000
Capsicum annuum	Sweet pepper	Hershenhorn *et al.*, 1996
Carum ajowan	Caraway	Qasem and Foy, 2007
Glycine max	Soybean	Qasem, 2006b
Lablab purpureus	Hyacinth bean	Acharya, 2012
Lathyrus ochrus	Ochrus vetch	
Medicago sativa	Alfalfa	
Phaseolus vulgaris	French bean	
Pisum sativum	Pea	
Cuminum cyminum	Cumin	Acharya, 2012
Foeniculum vulgare	Fennel	
Lens culinaris	Lentil	
Linum usitatissimum	Linseed	
Raphanus sativus	Radish	
Ricinus communis	Castor bean	Ghotbi *et al.*, 2012
Sesamum indicum	Sesame	

Vigna radiata	Green gram	Kleifeld *et al.*, 1994
Vigna unguiculata spp. *unguiculata*	Cowpea	http://www.uni-hohenheim.de/~www380/parasite/oro_path.htm

Orobanche cernua

Amaranthus spp.	Pigweeds	Rao, 1955
Bidens pilosa	Spanish needle	http://www.uni-hohenheim.de/~www380/parasite/oro_path.htm
Cajanus cajan	Pigeon pea	Krishnamurthy and Chandwani, 1975
Cannabis sativa	Hemp	Krishnamurthy and Rao, 1976; Krishnamurthy *et al.*, 1977
Capsicum annuum	Sweet chillies	Cited by Kasasian, 1971; http://www.uni-hohenheim.de/~www380/parasite/oro_path.htm
Cicer arientinum	Chickpeas	Krishnamurthy and Rao, 1976; Krishnamurthy *et al.*, 1977
Cichorium intybus	Chicory	http://www.uni-hohenheim.de/~www380/parasite/oro_path.htm
Colocasia sp.	Elephant's-ear	Rao, 1955
Crotalaria juncea	Sunhemp	Dhanapal and Struik, 1996; Dhanapal *et al.*, 2001
Curcuma domestica	Turmeric	Rao, 1955
Glycine max	Soybean	Krishnamurthy and Rao, 1976; Krishnamurthy *et al.*, 1977
Gossypium spp.	Cotton	http://www.uni-
Guizotia abyssinica	Nugcoat buttons	hohenheim.de/~www380/parasite/oro_path.htm
Hibiscus sabdariffa	Hibiscus	Rao, 1955
Illacrotvloma uniflorum	Horsegram	http://www.uni-
Linum usitatissimum	Linseed	hohenheim.de/~www380/parasite/oro_path.htm
Macrotyloma uniflorum	horsegram	
Medicago sativa	Lucerne	Krishnamurthy and Rao, 1976; Krishnamurthy *et al.*, 1977
Pennisetum typhoides	Millet	Rao, 1955
Phaseolus aconitifolius	Moth bean	
Phaseolus aureus	Green gram	Dhanapal *et al.*, 2001
Ricinus communis	Castor bean	Rao, 1955; http://www.uni-hohenheim.de/~www380/parasite/oro_path.htm
Sesamum indicum	Sesame	Rao, 1955
Setaria indica	Foxtail millet	

Sinapis alba	Wild mustard	Cited by Kasasian, 1971
Solanum melongena	Eggplant	Rao, 1955
Sorghum sp.	Sorghum	Krishnamurthy and Chandwani, 1975; Krishnamurthy and Rao, 1976; Krishnamurthy *et al.*, 1977
Tridax procumbens	Buttons	http://www.uni-hohenheim.de/~www380/parasite/oro_path.htm
Vicia dasycarpa spp. *villosa*	Vetch	Linke *et al.*, 1991
Vigna acontifolia	Moth bean	http://www.uni-
Vigna mungo	Black gram	
Vigna radiata	Mung bean	
hohenheim.de/~www380/parasite/oro_path.htmKrishnamurthy and Rao, 1976; Krishnamurthy *et al.*, 1977Dhanapal and Struik, 1996		
Setaria italica	Foxtail millet	Ye *et al.*, 2017
Orobanche crenata		
Allium sativum	Garlic	Hassan, 1998
Astragalus boeticus	Vetch	Schnell *et al.*, 1994
Brassica rapa	Turnip	Al-Menoufi and Adam, 1996
Capsicum annuum	Pepper	
Coriandrum sativum	Coriander	Al-Menoufi and Adam, 1996; Zemrag and Bajja, 2001
Glycine max	Soybean	Schnell *et al.,* 1994
Hedysarum coronarium	Sulla	
Helianthus annuus	Sunflower	http://www.uni-hohenheim.de/~www380/parasite/oro_path.htm
Hordeum vulgare	Barley	Kasasian, 1973; Linke *et al.*, 1991; http://www.uni-hohenheim.de/~www380/parasite/oro_path.htm
Lablab purpureus	Hyacinth bean	Schnell *et al.,* 1994
Lathyrus ochrus	Ochrus vetch	
Linum usitatissimum	Linseed	Khalaf, 1992; Abou-Salama, 1995
Lupinus termis	Lupine	Al-Menoufi and Adam, 1996
Phaseolus vulgaris	French bean	Schnell *et al.,* 1994
Pisum sativum	Pea	Hassan, 1998
Saccharum officinarum	Sugarcane	Abou-Salama, 1995
Sesamum indicum	Sesame	Al-Menoufi, 1991
Trifolium alexandrinum	Berseem	Schnell *et al.,* 1994; Al-Menoufi and Adam, 1996
Trigonella foenum-graecum	Fenugreek	Al-Menoufi and Adam, 1996; Zemrag and Bajja, 2001
Vicia dasycarpa spp. *villosa*	Vetch	Kasasian, 1973; Linke *et al.*, 1991

Schnell *et al.*, 1994

Vicia narbonensis	Narbonne vetch	
Vigna radiata	Green gram	
Vigna unguiculata	Cowpea	
Orobanche minor		
Allium sativum	Garlic	Hassan, 1998
Arabidopsis thaliana	Thale cress	Goldwasser *et al.*, 2000
Linum usitatissimum	Linseed	Brown *et al.*, 1952
Pisum sativum	Pea	Hassan, 1998
Sorghum bicolor	Sorghum	Brown *et al.*, 1952
Zea mays	Corn	

Source: modified after Qasem (2006a)

metabolites are known as allelochemicals (Qasem, 2006b; Cimmino *et al.*, 2015). In this regard, agronomic practices such as the use of allelopathic species in intercropping or cover crops or the use of direct seedling over residues of allelopathic species incorporate the principle of allelopathy exerted by molecules exuded from roots or released by crop residues to control broomrapes. In addition, the isolation of natural substances from root exudates of plants with the potential to inhibit broomrape development opens the door to design new herbicides based on natural and benign sources. Ryecyanatines A and B and ryecarbonitrilines A and B, the first new substituted cyanatophenol, substituted cyanatobenzo (1, 3) dioxole, and the latter two new substituted benzo (1, 3) dioxolecarbonitriles were isolated from rye (*Secale cereale* L.) root exudates. They were characterized as 4-cyanato-2-methoxyphenol, 2-cyanato-benzo (1, 3) dioxole, 2-methoxybenzo (1, 3) dioxole-5-carbonitrile, and benzo (1, 3) dioxole-2-carbonitrile by spectroscopic (essentially NMR and HRESI MS spectra) methods. These compounds were investigated for allelopathic activity on *Orobanche* germination and development. Ryecarbonitriline A stimulated seed germination of *Orobanche cumana* which can be considered suicidal since rye is not a host and ryecyanatine A promotes a rapid radicle growth cessation of *O. cumana*, *Orobanche crenata* and *Orobanche minor* with the promotion of a layer of papillae at the radicle tip in *O. cumana* and *O. crenata* disturbing the contact of the parasite to the host. Ryecarbonitriline B also showed a similareffect but with lower action compared with ryecyanatine A and mainly specific to *O. cumana* (Cimmino *et al.*, 2015).

In general, the use of trap crops may effectively reduce *Orobanche* seed bank, but to a certain extent these crops could be used as a part of integrated *Orobanche* control approach. Crop rotation alone will not solve the problem but it can contribute to reduce the infestation level to a degree that renders control methods like hand pulling. Good results with effective trap crops may not be efficient, possibly not practicable for ecological/economical reasons. Development of the parasite on weeds needs to be eliminated, while important trap crops to reduce the seed bank of different parasitic weeds are not available. The success of trap crops in exhausting soil seed bank of the parasite may be highly questionable under severe infestation or may require a fairly long period for effective control which may not be economically feasible under certain growing systems. However, Qasem (2006a) reviewed species reported as traps of different *Orobanche* spp. and other species of potential uses in this regard.

Catch Species

Catch crops are host plants (possible fodder harvest or green manure) that induce parasite seed germination and get parasitized (Fig.1.1.3). These crops are true hosts infected or enhanced germination and attachment but hinder parasite development by mechanical, physiological, or chemical factors. These must be sacrified, harvested, ploughed or destroyed after 6-8 weeks, mostly at vegetative and before the parasite emerge and produce seeds or latest before blooming of the parasite. It is safe and fast and can efficiently reduce the parasitic weed seed bank in the soil. This method can exhaust the parasite seed bank in the soil and may be effective in the long run. Catch crops are likely to be used as preceding crops; they have to be closely planted. However, this control method is costly due to additional labor, is not usable if the growing season is short, and needs good mechanization because of possible loss of the growing period. Cultivation of all hosts should be easy and inexpensive, germination stimulation should be high, their elimination is not problematic, and they are as high in yield as fodder or green manure. Acharya *et al.*, (2002) evaluated the effectiveness of toria (*Brassica campestris* var. *toria*) plant as a catch crop for the reduction of *O. aegyptiaca* seed bank. They found that two successive crops of toria reduced *O. aegyptiaca* seed bank by 20.9% and 26.2% for both crops, respectively. Optimum density of toria plants required for significant reduction of *O. aegyptiaca* seed bank was about 140 plants/m^2.

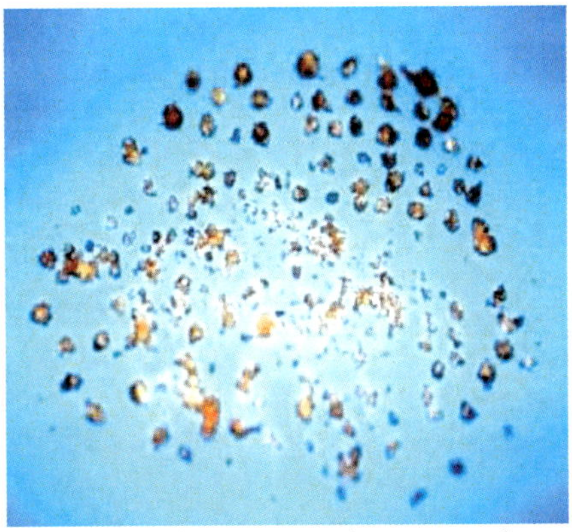

Fig. 1.1.3. Catch/trap species showing tubercles in a large number on roots but arrested growth

Intercropping with Trap or Catch Species

Intercropping was suggested by Al-Menoufi (1991, 1992) Al-Menoufi and Adam (1996), who reported a significant reduction in *O. crenata* infestation in broad bean intercropped with *Trigonella foenum-graecum*. In a later study, Al-Menoufi and Adam (1998) concluded the possible use of *Trigonella foenum-graecum*, *Lupinus termis*, *Coriandrum sativum* and *Brassica rapa* in intercropping with broad bean or tomato to reduce the infection rate of *Orobanche* attached to the host root systems. Many of these crops, however, were found attacked by *O. ramosa* (Qasem and Foy, 2007). Abbes *et al*., (2019) reported that intercropping with fenugreek (*Trigonella foenum-graecum*) enhanced seed yield and reduced *Orobanche foetida* infestation in broadbean (*Vicia faba*). Two years of field experiments and further investigation by mini-rhizotron and pot experiments showed a decrease of *O. crenata* infection due to an allelopathic interference on the parasitic life cycle at the level of germination. Inhibition of *O. crenata* seed germination by allelochemicals released by fenugreek roots is suggested to reduce parasite infection (Fernández-Aparicio *et al*., 2008b). The use of various mixtures of plant species is considered a possible strategy to affect *Orobanche*-host interaction by combining certain host species with different degrees of susceptibility (Bouhatous and Jacquard, 1994). Experimentation in Egypt reported a significant reduction of *O. crenata* infection on broadbean and pea inter-cropped with berseem clover. Mini-rhizotron experiments demonstrated reduction in *O. crenata* infection on pea, lentil and chickpea (Fernández-Aparicio *et al*., 2010a).

Intercropping system was also followed for *O. crenata* and *O. minor* control in pea. Garlic in mixture with peas resulted in best root system of peas and least *Orobanche* emergence, while higher seed yield was obtained when peas was mixed with onion. Complete elimination of *Orobanche* emergence was achieved in mixture with black cumin (Hassan, 1998). Maximum percentage of *O. cernua* germination was obtained when the parasite seeds were exposed to greengram, sunhemp and sesamum crops followed by blackgram and sunflower crops both in incubator and glasshouse experiments (Dhanapal *et al*., 1998). *Arabidopsis thaliana* was found to induce seed germination of *O. aegyptiaca*, *O. minor* and *O. ramosa* at a rate of 87, 72, and 67% of maximum seed germination, respectively (Goldwasser *et al*., 2000). Rodriguez-Ojeda *et al*., (2001) reported root exudates of corn, sorghum, millet, cotton, rice, eggplant and cauliflower to induce seed germination of *O. cumana* under laboratory conditions while none of these crops was infected with this parasite, suggesting that some could be used as trapping species. Certain weeds including *Avena sterilis, Conyza canadensis, Ammi majus, Datura stramonium, Cichorium endivia, Anagallis arvensis* and *Solanum nigrum* were mentioned to support *O. ramosa* attachment, but different resistant levels were observed and led later to death of the parasite. The authors suggested using these weeds as trap species for *O. ramosa* control (Boulet *et al.*, 2001). Qasem and Foy (2007) screened a large number of weed species for infection with *Orobanche ramosa* and found that many of these were heavily infected and others were nonhosts. However, a wide variation in tomato infestation was observed when the crop was planted after these weeds. *Orobanche crenata* infection on broadbean and pea was reduced when these host crops were intercropped with oat. The number of *O. crenata* shoots per host plant decreased as the proportion of oats increased in the intercrop. Pot and rhizotron experiments confirmed the reduction of infection in faba bean intercropped with cereals. The authors suggested that inhibition of *O. crenata* seed germination by allelochemicals released from cereal roots is the mechanism for reduction of *O. crenata* infection (Fernández-Aparicio *et al.*, 2007). Fernández-Aparicio *et al.* (2013) reported inhibition of *O. crenata* seed germination and radicle growth by cereals allelochemicals. In other studies Fernández-Aparicio *et al.* (2008a, b, 2010a) reported that intercrops with oat (*Avena sativa* L.), fenugreek (*Trigonella foenum-graecum* L.), or berseem clover (*Trifolium alexandrinum* L.) can reduce *O. crenata* infection on legumes because of allelopathy influence. Parasite seed germination was inhibited in the presence of oat or fenugreek roots, suggesting the release of toxic substances inhibited parasite seed germination, which was confirmed later. Trigoxazonane in fenugreek (Evidente *et al.* 2007) or benzoxazolinones (Fernández-Aparicio, unpublished) from oat root exudates may be responsible for the inhibition obtained.

Intercropping of canola with wheat could significantly reduce *Phelipanche aegyptiaca* growth depending on the type of wheat genotype. TRI565 and TRI12911, TRI15593, TRI18664, TRI19652, TRI17606, TRI19322, and TRI7259 genotypes showed strong inhibitory effects and can be considered as potential allelopathic genotypes for parasite suppression. The inhibitory potential of wild wheat genotypes was found stronger than cultivated wheat genotypes (Razavifar *et al.*, 2017). However, the success of trap crops in the exhausting soil seed bank of the parasite may be highly questionable under severe infestation or require a fairly long period for effective control, which may not be economically feasible under certain growing systems.

Plant Residues and Soil Cover

Surface applied and pre-plant incorporated wheat and barley straw residues used as mulch at different rates significantly reduced *O. ramosa* infestation and growth in potatoes (Haidar *et al.*, 1995). The use of olive cake in parasitic weeds (*Orobanche* spp.) control on different crops has also been reported (Ghosheh *et al.*, 1999, 2006; Aybeke, 2016, Disciglio *et al.*, 2016; 2018). Different ratios of soil-olive cake mixture reduced germination and infection of different *Orobanche* spp. on bean, peas, and tomato in pot experiments (Ghosheh *et al.*, 1999). Peas were not infected with *O. crenata* in the olive cake–containing media at any parasite inoculation density, while the only sporadic infection was detected on faba bean and tomato in media inoculated by *O. crenata* and *O. lavandulacea* on both crops, respectively. In a more recent study, Aybeke (2016) tested the effect of olive cake and different composting materials on *Orobanche cernua* infestation to sunflower. The author found that olive cake achieved success and was suggested to improve and provide regular plant development in arid lands of intense *Orobanche* infestation since fairly reduced parasite infection. Disciglio *et al.*, (2016) tested 12 agronomical, chemical, biological and biotechnological strategies included two rates of olive-mill wastewater for *Phelipanche ramosa* (L.) Pomel control in processing tomato. Still, none of the treatments provided complete control of the parasite. However, in a later study, Disciglio *et al.*, (2018) reported that olive mill wastewater was effective against *Phelipanche ramosa* infestation and increased tomato productive parameters.

Qasem (2020a) reported effective control of *O. ramosa* in potted tomato using olive mill wastewater and olive cake, but olive mill wastewater was more selective and effective against *O. ramosa* than olive cake and of a high potential use for parasite control in tomatoes. Qasem and Foy (2001), reported high allelopathic activity of shoot residues of certain weed species on seed germination of *O. ramosa* seeds in tomatoes grown under glasshouse conditions, and Qasem (2002) reported inhibitory effects of dried shoot residues and root exudates of different weed and crop species to *O. ramosa* infestation on tomato plants.

Soil Mulch and Solarization

Soil solarization is another method of *Orobanche* control practiced in the absence of crop plants and thus is a nonselective method that works against all species of the parasite and other weeds and soil pest. It is a method of *Orobanche* disinfection by heating moist soil using clear polyethylene (PE) sheet and natural sunlight. It involves the covering of moist soil by a layer of polyethylene under high-temperature conditions during the summer season and is applied in certain geographical temperate regions but not for other relatively cool or low temperate sites. Plastic mulch was found effective in controlling *Orobanche* sp. (Tsybul'skaya and Skoklyuk, 1978). Solarization with clear and black polyethylene mulches was tested for controlling broomrapes in naturally infested fields or host vegetable crops (Abu-Irmaileh, 1991). Only soil solarization has eliminated *Orobanche* and the majority of weeds from the treated plots. The yield was increased from 7.8 t/ha of untreated control plots to 21.08 t/ha on plots treated with solarization (Abdalla and Dabrowski, 2000). Jacobsohn *et al.* (1980) reported that *O. aegyptiaca* could be almost completely controlled in solarized soil and subsequent work has confirmed early results and refined the conditions required for success. Soil solarization with plastic sheets could be an additional measure to reduce the *Orobanche* infestation in small plots. It was reported that *O. aegyptiaca* was controlled with solar heating using polyethylene clear film 0.03 mm thick left in place for 30-50 days in the summer (Cartia, 1985). However, the efficacy of this technique may be improved when combined with organic supplementation, with positive effects on the yield of greenhouse tomatoes. Solar heating followed by mulching was promising for the control of *O. ramosa*, and increased eggplant yield with glyphosate used at 40g/ha (Braun *et al.*, 1985). Braun *et al.*, (1987; 1988) obtained high-level of *O. ramosa* control in eggplant in Sudan with this method for at least 30 days; Sauerborn and Saxena (1987) obtained corresponding results against O. *crenata* in faba bean and lentil in Syria, 40 days' treatment lead to > 90% control in both crops; and similarly. Abdel-Rahim *et al.* (1988) controlled *O. crenata* in faba bean in Egypt by 10 weeks of treatment. Soil solarization reduced population and incidence of *Orobanche* (Sauerborn *et al.*, 1989). Black plastic mulch completely controlled *O. ramosa* and *O. aegyptiaca* in tomato and eggplant (Vouzounis and Americanos, 1998). Soil solarization was also effective in controlling *O. cernua* in tobacco (Meti and Hosmani, 1994), and *O. crenata* in broadbean (Abdalla, 1999).

Soil solarization alone killed *Orobanche* seeds at 0 cm depth, but in combination with chicken manure, *Orobanche* seeds were killed at all soil depths from 0-10 cm (Haidar and Sidahmed, 2000). Solarization for 2-6 weeks with or without chicken manure significantly reduced weed growth and infestation in cabbage. Pre-plant composting with fresh manure under plastic mulch in the planting rows causes *Orobanche* seeds to lose viability within six weeks and reduced *Orobanche ramosa* infestation on many vegetables. This practice could be a useful asset in high-value crops. A subsistent farmer can easily practice fermenting manure in the farm without much input and can aid a sustainable farming strategy (FAO, 2008). It has been mentioned that two consecutive years of soil solarization are needed under the Mediterranean conditions to completely eradicate broomrape seed banks in highly infested soil (Mauro *et al.*, 2015). Incidental benefits from soil solarization can include controlling other weed species, nematodes, and other soil-borne pathogens. However, the relatively high cost of this method can be partly compensated by the control of difficult pests such as nematodes, most other weeds, fungal and bacterial pathogens and improved nutrient status of the soil. Furthermore, over several years following treatment, there is a residual effect, resulting in increased crop yields. This method, however, is not applicable in general for economic reasons, but under certain conditions, where water is available for irrigation,

Mineral Fertilization/Organic Matter

The negative effects on *Orobanche* species are mainly due to nitrogen. *Orobanche* tends to be in association with less fertile soil conditions. High levels of chicken manure and fertilizer, especially nitrogen, have a suppressive effect. Nitrogen fertilizer reduced *O. crenata* (Kukula and Masri, 1984). Ammonium nitrate applied alone or in combination with potassium phosphate completely inhibited infection of tomato by *O. aegyptiaca* (Jain and Foy, 1987). Ammonium nitrate with potassium phosphate or ammonium phosphate alone was most effective in reducing parasitism and enhancing the growth of tomato plants in *O. aegyptiaca* infested fields (Mariam and Suwanketnikom, 2004). Low rates of ammonium sulphate reduced *O. crenata* infestation without having an adverse effect on *V. faba* (Jain and Foy, 1992). High levels of nitrogen reduced the infestation of *O. ramosa* in tobacco and tomato, but it also reduced tomato yield (Personal observations). When ammonium-N based liquid fertilizer was in direct contact with *O. crenata* it adversely affected parasite germination potential. The adverse effect was dependent on the rate of N-used (Eplee *et al.*, 1994). The number of emerged shoots and dry weight of *O. ramosa* in tomatoes were decreased with an increase in N level, while an increase in the level of K had an opposite effect (Demirkan and Nemli, 1994).

In vitro Orobanche emergence could be reduced by applying nitrogen, but only a few field studies confirmed the reduction of *Orobanche* infestation by nitrogen application. Ammonium sulfate (8mM) applied during the conditioning period reduced germination of *O. crenata*, 4mM ammonium sulfate strongly inhibited germination when applied after conditioning during the germination phase. Urea at 8 mM reduced germination to only a limited extent (Van Hezewijk and Verkleu, 1996). Radicle lengths of *O. aegyptiaca* and *O. ramosa* were inhibited by ammonium nitrate and ammonium chloride than by potassium nitrate. Both parasite species were different in their sensitivity to nutrients (Nandula *et al.*, 1996). Ammonium form was more inhibitory to *Orobanche* spp. than nitrate (Westwood and Foy, 1999). The input of mineral fertilizer to the cropping system under consideration is rather low, probably just sufficient to replace the nutrient loss by harvest. For a considerable reduction in *Orobanche* infestation, the N-supply by the input of mineral fertilizers is by far not sufficient, and for economic reasons, it is not very promising to increase the input of mineral fertilizers to reduce *Orobanche* infestation. Germination of *O. aegyptiaca* and *O. oxyloba* was strongly inhibited by copper and vanadium, while boron inhibited radicle elongation of *O. oxyloba* at 25 ug/litre (Zaitoun *et al.*, 1996) and foliar spray of 0.5% boric acid 60 days after *V. faba* sowing completely controlled *O. crenata* (Hassan and Farrag, 1982). Elongation of *O. aegyptiaca* and *O. oxyloba* was inhibited by Zn and Mo at all concentrations used (Zaitoun *et al.*, 1996).

Metham sodium was also found to affect *O. aegyptiaca* germination (Goldwasser *et al.*, 1995)

Flooding

Soil flooding for an extended period of time (weeks), killed parasite seeds in the soil (Goldwasser and Rodenburg, 2014). Under rainfed agriculture, water shortage usually could not permit this method. Regular irrigation will support the host to compensate for the attack of the parasite to a certain extent. However, water lodging inhibited *Orobanche* emergence when applied for more than 6 weeks (Mohamed-Ahmed and Drennan, 1994). The effect of flooding, however, may work through denying oxygen required for the small embryo in the seed and elevating the toxic CO2 level in the seed surroundings.

Resistant and Toleranties

Research on crop resistance to *Orobanche* has not received much attention (Cubero, 1991). However, good levels of resistance have been found in several host/parasite systems as the sunflower to *O. cernua*, broadbean to *O. crenata* and common vetch to *O. crenata*, eggplant to *O. aegyptiaca* and several cucurbits to *O. ramosa* and *O. aegyptiaca*. Observations on the collection of peas and chickpeas also indicate the existence of strong levels of resistance (Alonso, 1998).

Resistant/tolerant varieties to *Orobanche* have been developed in several crops and used for some years. The most outstanding example has been the development of sunflower varieties resistant to *O. cernua/cumana*. Unfortunately, this resistance has often been overcome by new, more virulent 'races' of the parasite in many countries across the Mediterranean region, Eastern Europe and the former Soviet Union. Two cultivars of faba bean (Giza 429 and Giza 674) with a good level of resistance to *O. crenata* have been released in the middle- and upper-Egypt (Khalil *et al.*, 1993). Furthermore, a new faba bean genotype (X-843) resistance to *O. crenata* derived from Giza 402 was reported to have a good yield performance and was recommended for release in north Egypt (Saber *et al.*, 1999). A well-adapted, high-yielding faba bean cultivar 'Baraca' was developed in Spain under field conditions, with a high level of resistance to *O. crenata* (Cubero*et al.*, 1994). Alonso (1998) has intensively reviewed the most significant results achieved in the breeding for resistance to other *Orobanche* spp. Most of the reports on other crops referred to tolerance and moderate levels of resistance and were unable to detect a reliable high level of resistance. This has been the case for the following parasite-host relationship: *Orobanche aegyptiaca* and *O. ramosa* (tomato), *O. ramosa* (hemp), *O. crenata* (lentils, chickpea and pea), *O. aegyptiaca* (cucurbits), and *O. aegyptiaca*, *O. ramosa* and *O. cernua* (tobacco). In some cases, complete resistance under field conditions of some vetch lines (473A) (Gil *et al.*, 1987; Goldwasser *et al.* 1997), and tomato lines (PZU-11) (Foy *et al.*, 1988) was confirmed. Nevertheless, under artificial inoculation with a very high dose of *Orobanche,* even the field resistant lines were attacked by broomrape.

In a more recent study, Sauerborn *et al.* (2002) induced resistance in sunflower against *O. cumana* using the synthetic chemical benzo (1,2,3) thiadiazole-7-carbothioic acid S-methyl ester (BTH), the active ingredient of Bion®, under controlled growth chamber conditions. However, the application of BTH failed to induce resistance in sorghum and maize against *Striga* (Zain, 2002).

Among novel control methods, the breeding of resistant crops is partially utilized in Greece. In particular, 'resistant hybrids' of sunflower result in satisfactory control of the parasite. Sunflower is the third most affected crop by *Orobanche,* and it is mainly grown in the northeast part of Greece. It is noticeable in this area that there is no application of herbicides to control broomrape; the parasite is effectively controlled using two resistant hybrids, Favorite (Ypsilon S.A.) and Odil (Pioneer) (Syka and Economou, 2005).

Mechanism of Host Resistance to Parasites

It has been reported that once the parasite contacts and adheres to a host root, lytic enzymes such as pectin methyl esterase and polygalacturonase act to loosen the adhesion between host cells (Ben-Hod *et al.*, 1993; Joel and Losner-Goshen, 1994). In response to that host plants may show opposite reaction to prevent parasite attachment to its tissue surface. Antonova and Borg (1996) showed that the extracellular peroxidase in *O. cumana* races C and D reacts with phenolic compounds which are lignin precursor of the host sunflower resulting in host resistance due to the formation of lignin layers in sunflower plants possessing the Or3 gene for resistance. However, the absence of the extracellular peroxidase in parasite race D prevents lignin formation and enables the parasite to attach to the host vascular system. Wegmann *et al.*, (1991) reported that phytoalexines are almost certainly involved in the resistance response to *Orobanche cumana* cv. 81-14 showed twice as much scopoletin in its roots as did a susceptible line Giganta sunflower. These researchers however, couldn't confirm that phytoalexines played a role in the resistance of broad bean. Ish-Shalom-Gordon *et al.* (1993) noticed presence of phenolic materials in roots of resistant sunflower in response to haustorial penetration. Secretion of inhibitors (phytoalexins) by the resistant sunflower cultivars to *Orobanche cumana* infestation has been also suggested by Jorrin *et al.* (1996). Eizenberg *et al.* (2001) postulated that under summer conditions *O. cumana* penetrates and forms attachments on roots of the resistant cultivar indicating possible involvement of phytoalexines in the resistance reaction. These researchers reported that methanolic crude extract exhibited the strongest phytotoxic activity. They assumed that the resistance mechanism(s) of the resistant Amber cultivar is associated with phytotoxic substances produced by the host, but the amount produced at low temperatures was not sufficient to prevent parasitism. They further hypothesized that these phytotoxic substances inhibit broomrape development after penetration and establishment in host roots.

Different weed species showed various resistant levels including endodermic barrier, encapsulation layer, thickening of host vessels, vessels occlusion and cell wall polyphenolic impregnation in *Anagallis arvensis* and *Solanum nigrum* (Boulet *et al.*, 2001) that lead to the death of *Orobanche ramosa*. Field screening has shown very high levels of resistance to *Orobanche crenata* in *Medicago truncatula* a forage species receives the researchers' interest as a model plant for structural and functional genomic studies that can be used to identify agronomically important genes in legumes (Rodriguez-Conde *et al.*, 2004).

Other workers reported vessels of host plants to be completely filled up by phenolic compounds, gel or gum-like substances in combinations of LR1 sunflower/*O. cumana* and Palaiseau carrot/*O. ramosa*. Cell wall polyphenolic impregnation was also noted in *Helianthus rigidus*/*O. cumana* and *Anagallis arvensis*/*O. ramosa* (Labrousse *et al.*, 2001). Sillero *et al.* (2001b) reported that wild legumes of *Vicia*, *Pisum*, *Lathyrus*, *Cicer* and *Lens* all induced a lower germination of *O. crenata* seeds than the susceptible control crops being particularly low in *Cicer* and *Lathyrus*. The authors suggested that low germination of the parasite is due to low excretion of germination stimulants instead of excretion of germination inhibitors. However, it is well established that wild varieties are good sources of allelochemicals and have a greater effect through this mechanism than cultivated species or cultivars (Rice 1974; Putnam and Duke, 1978; Lovett *et al.*, 1982).

Literature on the resistance of different crop species to *Orobanche* is summarized below.

Broadbean (*O. crenata*)

An accession from the paucijuga group (VF 172) was reported to have a certain level of resistance (Cubero, 1973). A study by Zaitoun and Sabra (2000) confirmed that the new broadbean resistant varieties (Giza 667 and Giza 843) were less infected by *O. crenata* and the host resistance is related to the level of phenolic compounds, especially in its roots. The degree of variation in susceptibility within one species is less than between species. Lines with partial resistance were also identified (http://www.uni-hohenheim.de/~www380/parasite/oro_path.htm) including Line 561 (cross between Giza 402 X Seville giant); lines 667 and 674 ("Giza 402" as one parent, composite crosses), Various lines (F8 generation) in Spain, having "Giza 402" as a parent Paucijuga-types are usually more resistant. The line F402 was the first reported to show a high resistance and acceptable agronomical characters, and thus Giza 402 variety was derived from this line. In turn, the line TVF 1071 was derived from Giza 402 and became better resistant than the mother variety (Cubero and Hernandez, 1991), and later the cultivar Baraca was obtained and regarded as a high yielding and a very well adapted cultivar (Cubero *et al.*, 1992).

Resistance of several faba bean varieties against more than two *Orobanche*/*Phelipanche* spp. was also confirmed under field conditions (Trabelsi *et al.*, 2015, 2016). Giza 402 could also be regarded as tolerant to *O. crenata* since the parasite had no significant effect on its pod and seed production (Zaitoun *et al.*, 1991). Among other reported resistant lines were 18009, 18035 and 18105 (ICARDA, 1992). However, a number of faba bean cultivars with resistance to *O. crenata* have been released, all of which use an Egyptian line Giza 402 as the main donor of resistance (Fernández-Aparicio *et al.*, 2012; Rubiales *et al.*, 2016).

A germplasm collection of 483 *V. faba* accessions was screened by Rubiales *et al.* (2014) for resistance to *O. crenata* under field conditions. The most resistant 37 germplasm were re-tested in different countries and showed significant levels of stable resistance across environments. Seven quantitative trait loci for *O. crenata* resistance and three for *O. foetida* were identified using the faba bean. Co-localization of quantitative trait loci for *O. crenata* and *O. foetida* resistance in chromosome V confirms a common resistance against both parasites (Guti'errez *et al.,* 2013). Cultivar Baraca and the accessions V-1268, V-1302, V-1301, V-268, V-231, V-319 and V-1272 were the most resistant, stable across environments, and resistant to both *O. crenata* and *O. foetida*. Compared to the faba bean local variety, Hashenge, Walki and Moti varieties reduced the incidence of *O. crenata* infestation by 74, 11 and 10%, respectively. Similarly, the severity of *O. crenata* infestation was reduced by 79% on Hashenge variety, 30% on Walki and 16% on Moti variety compared to the negative check (no bio-control) (Aregawi, 2017).

A galactosyl-sucrose trisaccharide, planteose, was found to be a required storage carbohydrate for the early stage of germination of *O. minor* through a metabolomics analysis of germinating seeds of the parasite.

Nojirimycin bisulfite inhibited the germination of *O. minor* seeds by blocking the supply of glucose from planteose and sucrose, suggesting that a galactosidase inhibitor such as nojirimycin bisulfite is a lead chemical to develop an *O. minor*-specific herbicide (Wakabayashi *et al.*, 2015). A proteome analysis of the pea parasitized by *O. crenata* (Castillejo *et al.*, 2012), a metabolome analysis of *P. aegyptiaca* at several growth stages (Nativ *et al.*, 2017) and a transcriptome analysis of several parasitic weeds (Westwood *et al*, 2012) might lead to the identification of new targets for the specific control of parasitic weeds.

Lentil (O. crenata)

In a laboratory screening study, the ILL 5748 line showed 50% fewer parasite infections and the ILL 8. line is not less susceptible, but as it is suitable for late planting, it escapes the attack. The higher levels of resistance were observed in the accessions of wild lentil *Lens ervoides, Lens odemensis* and *Lens orientalis*. Resistance of selected accessions proved to be mainly due to early hampered tubercle formation, with no observed necrosis of established tubercles (Fernández-Aparicio *et al*, 2009a). In general, low intraspecific variation in susceptibility to *Orobanche* exists in lentils compared to broadbean. Screening wild lentil species, therefore, might be promising and necessary.

Chickpea (O. crenata)

Indication of resistance was found in ILL 3279 and some other lines. Chickpea, in general, is much less infested (by a factor of about 8) than broadbeans. Quite some intraspecific variation is to be expected. Variety ILC-280 was also mentioned as a resistant one with a substantial amount of phytoalexins (Wegmann *et al.*, 1991). Recently, three genotypes namely, FLIP 98- 22C, Nayer, and Beja 1, were reported to show partial resistance to *O. foetida* (Nefzi *et al.*, 2016). Their resistance was ascribed to incompatibility.

Sesame (Sesamum indicum) (Phelipanche aegyptiaca)

Five resistant sesame genotypes with an important trait were identified. *3KHOLSDQFKH aegyptiaca* formed tubercles on their roots, but no parasite shoots emerged. Such sesame genotypes could be used as both trap crops and as breeding sources for the development of resistant genotypes (Teimouri *et al.*, 2016).

Field pea (Pisum sativum L.) (O. crenata)

A low *O. crenata* infestation in the field was noticed in "Facima" and in the ICARDA accession "290". Rubiales *et al.* (2005) screened 575 accessions of *Pisum sativum* for resistance to *O. crenata* under field condition, and no complete resistance was found, but varying levels of resistance were identified and confirmed between accessions. A highly *O. crenata*-resistant pea line, named ROR12, was identified (Bardaro *et al.*, 2016; Pavan *et al.*, 2016). The resistance mechanism may be due to low release or production of germination stimulants but this does not affect the host yield of the resistant line.

Fodder legumes (O. crenata/O. aegyptiaca)

Resistance was found in Spain with *Vicia sativa* lines, including lines 473A, 485B, 486A and 417B (Gil *et al.*, 1987). At ICARDA, Syria, Linke *et al.* (1993) reported high interspecific and intraspecific variation between five lines of *Vicia narbonensis* L., *V. sativa* L. and *V. villosa* ssp. *dasycarpa* (Ten.) Cavill. *Lathyrus sativus* L., *Lathyrus ochrus* (L.) D.C. and *Lathyrus cicera* L. for resistance to *O. crenata. Lathyrus ochrus* lines were free of emerged *O. crenata* shoots, while those of *L. sativus* and *L. cicera* were highly susceptible to the parasite. Intraspecific variation was high within *V. narbonensis* and *V. sativa,* with several lines being free of *Orobanche* parasitism, whereas others were highly susceptible. Resistance was found in *V. sativa* lines 715 and 2541, *V. narbonensis* line 578, *V. villosa* ssp. *dasycarpa*, *Lathyrus ochrus*, and *L. sativus* line 201. The cultivar Mezquita was more resistant than PI-183723 against the two populations of *O. crenata* studied, and also against *O. aegyptiaca* (Fernández-Aparicio *et al.*, 2009b). Sadot cultivar developed from *Vicia atropurpurea* was reported as highly resistant to both *Orobanche* species (Goldwasser *et al.*, 1996; 1997).

Grass pea (Lathyrus sativus)
Resistance to *O. crenata* has been detected in wild peas (*Lathyrus cicera*) which was due to a low induction of germination, incompatibility and escape parasite attack due to precocity or small root biomass (Fernández-Aparicio and Rubiales, 2010). Screening studies on the responses of 52 grass pea accessions to *O. crenata* under field conditions revealed that early and very late-maturing accessions were less infected (Fernández-Aparicio *et al.*, 2012). Nine more accessions would have true genetic resistance that was not dependent on the short and long growth period (Fernández-Aparicio *et al.*, 2012).

Parsley (O. crenata and O. aegyptiaca)
Differences between different varieties of parsley in their susceptibility or tolerance to *O. crenata* and *O. aegyptiaca* have been reported (Goldwasser and Kleifeld, 2002). Garland and Garbo parsley varieties were tolerant, while Darki, Extra triple and Kad erava were susceptible to both parasitic species.

Oilseed rape (Brassica napus)
The cultivars Darmor, Campo and Shakira, had resistance based on a low induction of germination, and the resistance of five Darmor, Campo, Adrianna, Expert and Shakira cultivars was ascribed to incompatibility (Gauthier *et al.* 2012).

Cucurbites (O. aegyptiaca)
Summer varieties of musk melon, water melon and cucumber were most resistant to *Orobanche* spp. due to the short growing season. Kuzibai was the most resistant water melon cultivar (Mukumov and Faizieva, 1977). *Orobanche aegyptiaca* on *Cucurbita pepo,* tracheids were formed in the haustoria but didn't penetrate the vessels of the host plant, indicating considerable resistance to the parasite. In musk melon resistant varieties cv. Kukkalapush and Umyrvaki, the haustoria did not form tracheids and derived nutrients from parenchyma cells. In resistant cucumber variety cv. Margelanskii, *Orobanche* formed single vessels in the haustoria but these failed to penetrate vessels of the host plant and growth of the parasite was greatly stunted (Kabulov and Kabulova, 1977).

Sunflower (O. cernua/O. cumana)
Races of broomrape have been identified and sources of resistance appear to have stabilized sunflower production in certain European countries, but racial evolution of broomrape has been very rapid during the past few years, and all resistance has been overcome. Vranceanu *et al.* (1980) used a set of sunflower genotypes and identified that each of them carries one single major gene of resistance (Or1 to Or5) and that these genes control races A to E of *O. cumana*, respectively.

In Spain, three sets of cultivated plant material were tested, including 160 USDA breeding lines. Out of these lines, 5% were completely resistant and 19% segregated for resistance. These lines trace back to crosses of two USDA restorer lines with Russian, Turkish and Romanian hybrids. Also, the origin of PI accessions from the USDA Plant Introduction Station at Ames, IA, that segregated for resistance were primarily derived from the former USSR and Romania (Dominguez *et al.*, 1993).

Varieties with various degree of resistance are available, but mainly in oil sunflowers. The commercial sunflower cultivar M15 from Argentina showed complete resistance to race E of *O. cumana* (Miladinovic *et al.*, 2012). Line no. 16693 cv. Armavirskii 3497 was completely resistant to different races of *O. cumana*, and the inbred line AB-VL-8, developed from interspecific hybridization with the rough sunflower (*Helianthus divaricatus*) was a successful example of full resistance to races higher than F (Cvejic *et al.*, 2012; Imerovski *et al.*, 2016). Successful identification of quantitative trait loci (QTLs) for *O. cumana* resistance in the sunflower has been reported (Akhtouch *et al.*, 2016; Louarn *et al.*, 2016) and utilization of QTLs in the breeding process is expected.

In *O. cumana*-resistant sunflower genotypes, up-regulation of phenylalanine ammonia lyase (PAL) and defense genes or the PR5 gene was found (Sestacova *et al.*, 2016). Start cultivar was found to show 98% resistant to different races of *Orobanche* (Mamonov *et al.*, 1978) and Tubro cultivar showed resistance to *O. cumana* (Vranceanu *et al.*, 1990). One important source of resistance is *Helianthus tuberosus*. It was found that all perennial *Helianthus* spp. were resistant to different populations of *O. cernua* (Ruso *et al.,* 1994) and have been frequently used. In a more recent study on resistance among 23 wild species of perennial *Helianthus*, all were found resistant/immune to the two widely spread *Orobanche* isolates. *Helianthus gracilentus* and *H. muttallii* spp. *muttalli* showed resistance segregation (Ruso *et al.*, 1994, 1996; Dominguez *et al.*, 1996 and Sukno *et al*, 1996). *Helianthus exilis*, *H. anomalus* and *H. agrestris* were the only resistant annual species (Sukno *et al.*, 1996). Sunflower cultivars SW-501, RW-637, SW-501, Sunbred-254 and RW-637 were reported as resistant to *O. cumana* (Ish-Shalom-Gordon *et al.*, 1993) and R-41 inbred line was found resistant to *O. cumana* and conferred by two independent dominant genes (Dominguez, 1996). Sunflower inbred line KI-374 showed high resistance to *O. cernua* and *O. cumana* (Rodriguez-Ojeda *et al.*, 2001). Lines MX199 and MX655 were reported to have complete resistance to the most virulent race of *O. cumana* (Batchvarova *et al.*, 2001).

<u>*Eggplants*</u>

Recent reports on the resistance of this crop to *Orobanche* are lacking. However, results reported by Dalela and Mathur (1971) are still regarded as the only source of information on this crop. These authors mentioned that out of 128 cultivars tested for resistance to *O. cernua*, the highly resistant were the hybrid of Pusa Purple Long with Manjri Gota or with Nurki, Running King, Verma's Giant, D-12-2-66, DC-4-1-67 and E-147.

<u>*Tomato (O. ramosa/O. aegyptiaca)*</u>

Long time ago some resistance was reported from the Russian cultivars "Ora". cv. Gibrid-88 and cv. Bolgarskii which were the most tolerant to *Orobanche* infestation (Mukumov and Faizieva, 1977). Avdeev and Scherbinin (1983) reported some important resistant varieties to *O. cernua* including *Lycopersicon hirsutum*, *Lycopersicon pyriforme* and *Lycopersicon. pruniforme*. Avdeev and Shcherbinin (1977; 1978) found that UZ-1 as almost an immune line to *O. aegyptiaca* and a later study proved that PZU-11 line is resistant to the same parasitic species (Avdeev, 1986). However, conflict results were obtained by Cubero (1991) on the same line and Foy *et al.* (1988) were unable to confirm resistance of the same line. Tomato line CE-TO-32-2/82 was also reported as the most resistant to *Orobanche* infestation with only 3% infection (Sobrino Vesperinas, 1985). Kasrawi and Abu-Irmaileh (1989) reported LA 372; a wild accession of *Lycopersicon peruvianum* as a resistant line to *O. ramosa*. In more recent studies, high to moderate resistant to *O. ramosa* was obtained in the cultivars Tiny Tim, Acora, Cestor, Promodora, Orient, Red Alert and *Lycopersicon pimpinellifolium* LA 1478 with Tiny Tim showed the highest level of resistance (Qasem and Kasrawi, 1995). Kacan and Tursun (2012) reported several tomato cultivars as resistant to *Phelipanche aegyptiaca* in Turkey and Tokasi *et al.* (2014) reported certain other tomato varieties to resist the same parasite under field conditions in Iran. Avdeev (1998) reported that tomato var. Astrakhan and Bakhtemir were not infested by *O. aegyptiaca* and there was no effect on yield. Sl-ORT1 mutant of tomato was shown to be resistant to high concentrations of *P.*

aegyptiaca seeds, and to another three broomrape species: *Phelipanche ramosa*, *Orobanche cernua*, and *Orobanche crenata* (Dor *et al.*, 2010). The fast-neutron-mutagenized tomato mutant SL-ORT1 was found highly resistant to various *Phelipanche* and *Orobanche* spp. SL-ORT1 resistance results from its inability to produce and secrete natural germination stimulants to the rhizosphere (Dor *et al.*, 2011). Tomato cultivar 'Red setter tilling', developed in the genetic background of a cultivar 'Red setter' using a targeting-induced local lesion in genomes (TILLING) platform, was resistant to *Phelipanche ramosa* because of a reduction in the synthesis of the germination stimulant (Disciglio *et al.*, 2016).

Tobacco (*O. cernua*/ *O. crenata*/ *O. ramosa*)

Wide variations have been long reported between tobacco varieties, cultivars, and accessions in their susceptibility/tolerance/resistance to *Orobanche* species infestation (Racovitza, 1973). Moderate resistance was detected in different tobacco lines and cultivars (Vinogradov *et al.*, 1981; Sarychev *et al.*, 1982; Vinogradov *et al.*, 1984; Raju, 1996).

Among the tolerant and resistant lines already mentioned were Trebizond-285, Corojo, Florida-513, NS-2512, Hicks Resistant, Khatana, R X T, and Jamaica K-444 (Vinogradov *et al.*, 1981) and some degree of resistance to *O. ramosa* and *O. aegyptiaca* in some lines were expected. Buschmann *et al.* (2005) reported on differences in tobacco cultivars to *O. ramosa.* Populations' Dark air-cured tobacco cultivars were the least susceptible to broomrape populations. The tobacco genotypes, TB 22 and Kramograd NHH 659, did not show any susceptibility to *P. aegyptiaca* in pot experiments (Darvishzadeh, 2016).

Biological Methods

A number of biological, environmental and technological limitations, variability of field performance, cost of production and market size are the major constraints that potentially affect the economic feasibility of any given biological control product (Auld and Morin, 1995). However, below are the natural enemies that could be used against these parasites.

Insects

A large number of insects have been reported to attack *Orobanche* species, but most are not specific but rather polyphygeous. Tóth *et al.* (2005) conducted a survey of insects on broomrapes (*Orobanche* spp.) and reported 22 species belonging to 10 different families in Slovakia. Only two flies, *Phytomyza orobanchia* and *Chyliza extenuata*, and two moths, *Diaphora mendica* and *Celypha* spp. caused significant damage. *Chyliza extenuata* as a root feeder and *D. mendica* as a seed capsule feeder seem to be very promising.

Biological control using the restricted oligophagous fly *P. orobanchia* Kalt. (Diptera, Agromyzidae) (Fig. 1.1.4) has been successfully applied on a large scale in the former Soviet Union to control *O. cumana* Wallr. and *O. cernua* Loefl. (Kapralov, 1974, 1975).

The insect (Fig. 1.1.4) attacks seed capsules and cause a great reduction in seed production of the parasite ranging between 29-94% depending on *Orobanche* species. This insect, however, is influenced by different factors, including the presence of its natural enemies in the field and the agricultural practices, mainly the use of insecticides and tillage, which may limit its efficiency. Most important are larval and pupal parasitoids of the order Hymenoptera of 24 known species of seven families (Klein and Kroschel, 2002).

In addition, other insects have already been reported to attack *Orobanche* plants, including *Psila* species at the larval stage that attack stem of *O. minor* and *C. extenuate*. *Eulocastra argentisparsa*

Hampson, *Smicronyx* spp., *Ophiomyia strigalis* Spencer have also been reported on *Orobanche* in India and Africa, but are polyphageous (Kroschel *et al.* 1999; Traoré *et al.*, 1999; Klein and Kroschel, 2002). However, the phytophagous insects on *Orobanche,* including *P. orobanchia* occur in most *Orobanche* infested regions.

Fig. 1.1.4. *Phytomyza orobanchia*

Phytomyza is a monophagous fly 2.5-3 mm long, black and lays eggs (1-3) into the bud, flower or stem. It was reported to attack a wide range of *Orobanche* populations either on crops or natural in the wild (Rubiales *et al.*, 2001). The female can lay 180-200 eggs in its life time. The duration of the life cycle is 30–35 days. Depending on the climatic conditions and the vegetative stage of the host, 2-3 generations occur in Eastern Europe and 4-5 in North Africa and West Asia. *Phytomyza* hibernates as pupae in the underground parts of the stem and the fly emerges together with *Orobanche* plant. The larvae feed on the shoots and especially on the capsules of *Orobanche* spp. (Klein *et al.*, 1999). Depending on the insect population, they can seriously damage *Orobanche* and may reduce seed production by 30% under natural infestation in Syria and 89% in Egypt. Natural enemies can reduce the fly population considerably, 12 species in five families have been reported to parasitize *Phytomyza* (Eltayeb *et. al*., 2000).

At present, the only way to use *P. orobanchia* is to collect *Orobanche* infested with the fly's larvae and distribute these shoots in the field when *Orobanche* is expected to emerge. This technique was first developed in Eastern Europe for *O. cernua* control in sunflower and applied against *O. crenata* in Morocco. Preliminary field releases have shown more than 95% efficacy depending on the site, the host plant, and the *Orobanche* species. A natural reduction of the seed production between 11-79% was reported from several countries of the Mediterranean region, North and East Africa, and the Near East (Klein and Kroschel, 2002).

Phytomyza puparia also has been reported to feed on *Orobanche* seeds and damage the stalk of the parasite (Tsybul'skaya and Skoklyuk, 1978; Tsybul'skaya *et al.*, 1978). It has been reported on *O. crenata* (Hassanein *et al.*, 1998). The efficacy of *P. orobanchia* as a biological control agent of *Orobanche* is well established, and the fly is frequently occurring on *Orobanche* in different Eastern European countries. Under natural conditions, the seed production was reduced by about 41% on average.

Elzein *et al.* (1999) reported that *P. orobanchia* destroyed *O. ramosa* and *O. cernua* seed capsules by 81.4 and 71.7%, respectively, in tomato fields. Mass rearing of *Phytomyza* and the yearly release could be a possible approach of *Orobanche* control in the Mediterranean area but, it has to be considered to adapt the regular aphid control by insecticides to this approach. In general, *P. orobanchia* and *Smicronyx cyaneus* Gyll. (Coleoptera, Curculionidae) are of great interest in biocontrol of *Orobanche* spp. (Klein and Kroschel, 2002). The host range of *S. cyaneus* as another promising species needs more investigation.

Table1.1.4. Insects (Hymenoptera) parasitizing *Phytomyza orobanchia* (After Sauerborn, 1991; Eltayeb *et al.*, 2000).

Aphelinidae	**Pteromalidae**
Aphelinus chaonia Walk.	*Callitula bicolor* Spin.
Braconidae	*Cyrtogaster vulgaris* Walk.
Opius occulisus Tel.	*Eupteromalus micropterus* Lind.
Opius similoides	*Homoporus subniger* Walk.
Encyrtidae	*Sphegigaster flavicornis* Walk.
Thomsonisca typical Merc.	*Sphegigaster orobanchiae* Kurd.
Eulophidae	*Stenomalina gracilis* Walk.
Crataepiella carlinarium Szel et Erd.	*Thinodytes cyzicus* Walk.
Tetrastichus phytomyzae Kosr.	*Trichomalus* sp.

Bacteria

In pot and Petri- dish experiments, Mabrouk *et al.* (2007) found that inoculation of peas with two (P.SOM and P.1236) of the five strains induced a significant decrease in *O. crenata* seed germination and the number of tubercles on pea roots. Symptoms, including the non-penetration of the germinated seeds into pea roots followed by radical browning and death of the parasites, were observed in the presence of these inoculated pea plants. The hypothesis is that roots secrete toxic compounds related to *Rhizobium* inoculation. In other pot experiments, chickpea roots inoculated with *Rhizobium* sp. strain PchAZM reduced the total number of *O. foetida* by up to 90% (Mabrouk *et al.*, 2016). Bouraoui *et al.* (2016) reported that despite the high seed yield reduction of faba bean caused by *O. foetida*, the inoculated faba bean with Mat strain of rhizobia showed an average yield three-fold higher than the control.

Two *Rhizobium leguminosarum* strains (Mateur and Bouselem) were selected as potential inoculants to protect faba beans against *O. foetida* and to promote the host's growth in pot and hydroponic culture (rhizotron) experiments (Bouraoui *et al.,* 2012). Inoculation of chickpeas with *Rhizobium* strains significantly decreased *O. crenata* and *O. foetida* seed germination and the number of tubercles (Hemissi *et al.*, 2013; Mabrouk and Belhadj, 2014). An endophytic bacterium, *Pseudomonas* strain PhelS10, originating from tomato roots, suppressed *P. aegyptiaca* seed germination and reduced the number of the same parasitic species on tomato roots (Kruh *et al.*, 2017). Rubialis *et al.* (2018) reported an almost 80% reduction in *P. aegyptiaca* seed germination and a significant reduction in its parasitism was achieved due to the presence of a *Pseudomonas* sp. strain on tomato roots.

Fungi

A total sixty-two fungi species were isolated from different *Orobanche* species (Thomas *et al.*, 1999a). Eight of these showed pathogenicity to *Orobanche*. Four belong to the genus *Fusarium*, among which *Fusarium oxysporum* f. sp. *orthocera* (Thomas *et al.*, 1998, 1999a) *Fusarium arthrosporioides* (Amsellem *et al.*, 1999) thought to be developed as mycoherbicides. Babalola (2010) reported that *F. arthrosporioides* killed 56% of *O. aegyptiaca* tubercles infesting tomatoes when applied with cellulase. The fungus mycelia had no damage effects on tomato roots. Other reports mentioned 16 fungi species attack different *Orobanche* species, including *O. aegyptiaca, O. cernua, O. crenata* and *O. ramosa* but only 4 genera proved highly effective and selective to *Orobanche* these were *Ulocladium* species, *Fusarium* species (*F. oxysporum var. orthoceras; F. orobanchas*), and *Alternaria* sp. and *Rhizoctonia solani* which was effective against *O. ramosa* in tomato and reduced its infestation by 80%, *Ulocladium atrum* was effective as a biocontrol agent for *O. crenata* at high humidity.

Fusarium arthrosporioides E4a and *F. oxysporum* E1d and strains of bacteria were isolated from diseased, juvenile, *Orobanche* flower stalks. They are pathogenic to *O. aegyptiaca, O. crenata* and *O. ramosa* on most vegetable crops. In addition, several fungi species were also reported to be virulent on *Orobanche*, and mycoplasma-like organisms were detected in *O. cernua* sap stream (Singh and Joshi, 1992). Tóth *et al.*, (2005) reported that pathogens were observed basically on two wild broomrapes, *O. alba* and *O. flava.* These were similar on both broomrapes and included, *Alternaria alternata, A. tenuissima, Aspergillus niger, A. flavus, Cladosporium cladosporioides, Epicoccum nigrum, Fusarium oxysporum, F. sambucinum, F. solani, F. sporotrichioides, F. tabacinum, Rhizopus nigricans, Stemphylium botryosum, Penicillium* spp., *Phoma* spp., *Pythium* spp., and *Trichoderma* spp. The exceptions were *Aspergillus flavus, Cladosporium cladosporioides* and *Rhizopus nigricans* which were isolated only from *O. flava*. The most important pathogen was *F. oxysporum*.

Except for Russia, where good results were obtained by adding *F. oxysporum* into the planting holes of watermelon (Product F), no practical use could be made so far of these fungi. However, current investigations concentrate mainly on *F. oxysporum* f. sp. *orthoceras*, a pathogen with high potential to control *O. cumana* parasitizing sunflower. The fungus was reported to control *O. cumana* in sunflower by 90% and *O. ramosa* and *O. aegyptiaca* were also found susceptible (Bedi, 1994). In other studies, the fungus decreased *Orobanche* infestation to tobacco by 75.23% and increased crop yield by 80.5% (Mazaheri *et al.*, 1991). Under field conditions, the fungus at 10g/kg of soil resulted in more than 90% control of *O. cumana* (Bedi and Donchev, 1991) and was reported to control *O. cumana* by 74-90% in naturally infested fields (Bedi, 1992) and thus recommended for *O. cumana* control in sunflower (Bedi and Donchev, 1995; Thomas *et al.*, 1998; Bedi and Sauerborn, 1999). It was found effective at a concentration of 69 conidia/ml water (Thomas *et al.*, 1998).

Fusarium solani was reported to suppress the growth of *O. ramosa* in tomatoes (Gold *et al.*, 1979) and *Fusarium oxysporum* f. sp. *orthoceras* (Müller-Stöver *et al.*, 2004) is of a high potential candidate for broomrape control. Studies are focusing on the suitable mass production method and formulation for use under field conditions. The microconidia or air-dried chlamydospore were formulated as granules with wheat flour and kolin (Pesta) as rich biomass of *F. oxysporum* Schlecht f. sp. *orthoceras* (Appel and Wollenw.) were prepared and found effective against *O. cumana* in sunflower in the greenhouse (Müller-Stöver *et al.*, 2002) and reduced shoot emergence of the parasite by 64% (Müller-Stöver *et al.*, 2004). The combination of *F. oxysporum* Schlecht. f.sp. *orthoceras* with BTH Benzo (1,2,3) thiadiazole-7-carbothioic acid S-methyl ester), a product that induced resistance to *O. cumana* in sunflower, was highly effective and sometimes better than when *Fusarium* was used separately (Müller-Stöver *et al.*, 2005).

In pot experiments, application of the fungus granules into the potted-soil reduced number and dry weight of *Orobanche* shoots by more than 90%. *Orobanche* shoots treated with the conidial suspension of the fungus dead by 75% in two weeks (Müller-Stöver *et al.*, 2009).

Conidial suspension of two selected isolates of *F. oxysporum* significantly reduced germination, attachments, and tubercles of *O. crenata*. The microconidia and chlamydospores of both isolates were formulated as mycoherbicides encapsulated in wheat flour–kaolinmatrix formulations greatly reduced the number of emerged *Orobanche* shoots, their heights, attachments, and dry weights. Meanwhile, disease incidence and disease severity of emerged shoots were enhanced (Nemat Alla *et al.*, 2008). *Fusarium oxysporum* caused heavy reactive oxygen species (ROS) damage in *Orobanche* induced significant irrevocable genotoxic effects on the DNA of *Orobanche*, degraded protein metabolism and synthesis, and finally triggered apoptosis (Aybeke, 2017a). It has been reported that it causes heavy hormonal disorder, triggered only SA-mediated defense, and induced the accumulation of phenolic substances in *Orobanche* intensively. *Fusarium oxysporum* causes lethal physiological damage on *Orobanche* spp. (Aybeke, 2017b).

Other plant pathogens reported including, *Alternaria*, *Fusarium* and *Trichoderma* were also found infesting *O. crenata* in Egypt (Abdel-Kader *et al*, 1998). *Ulocladium atrum* infected and destroyed underground tubercles and emerged shoots of *O. crenata* at 20°C and RH 50-80% (Linke *et al.*, 1992). *Ulocladium atrum* (Preuss) Sacc. and *Ulocladium botrytis* Preuss have been found pathogenic to *O. crenata* (Linke *et al.*, 1992; Müller-Stöver and Kroschel, 2005). *Myrothecium verrucaria* (Alb. and Schwein.) Ditmar isolated from faba bean roots has been found to inhibit the germination of *O. crenata* seeds due to the macrocyclic trichothecene verrucarin A (El-Kassas *et al.*, 2005). *Myrothecium verrucaria* inhibited the germination of conditioned *O. crenata* seeds exposed to the germination stimulant, GR24. The inhibitor was identified as verrucarin A and infection of *V. faba* by *O. crenata* could be prevented by addition of spores of the fungus to infested soil by the parasite (El-Kassas *et al.*, 2005).

The application of mycoherbicides *Trichoderma harzianum* and *Trichoderma viride* followed by foliar spray with glyphosate was the most appropriate treatment. It resulted in the highest reduction in the parasitic parameters of *O. ramosa*, although a lesser increase in tomato yield compared to the individual mycoherbicide approaches was recorded. Soil treatment with isolates of *Trichoderma* spp. including *T. harzianum* and *T. viride* alone or soil treatment with fungal agents plus aerial spray of glyphosate (50 ppm) was effective in reducing infection of broomrapes (*O. crenata* and *O. ramosa*) and increasing yields of peas, faba bean, and tomatoes (Abdel-Kader and El-Mougy, 2009). This work developed an effective, applicable and cost-effective method acceptable for controlling *O. ramosa* in tomato field (Abdel-Kader and El-Mougy, 2007). *Trichoderma harzianum*, *T. harzianum* + *Rhizobium leguminosarum* and *R. leguminosarum* reduced *O. crenata* incidence in faba bean by 11.5%, 8.4% and 7.6%, respectively. The same treatments reduced *O. crenata* severity by 39, 48, and 35%, respectively.

Among the bio-inoculants, *T. harzianum* + *R. leguminosarum* treated plots had the highest grain yield (803.50 kg/ha), which was a 27% over the untreated control (Aregawi, 2017).
Selective active *Azotobacter chroococcum* and *Escherichia coli* transformants added to soil infested with *O. crenata* seeds showed promising results for parasite control on broadbean (Khalaf *et al.*, 1997 a, b).

Arbuscular mycorrhizal (AM) fungi root exudates had a negative impact on the germination of *O. cumana* induced by germination stimulants (Louarn *et al.*, 2012; 2016).
A similar effect could be obtained with AM spore exudates, establishing the fungal origin of at least part of the active compounds (Louarn *et al.*, 2012). Production and exudation of strigolactone were significantly reduced by Arbuscular mycorrhizal fungi symbiosis in tomatoes, and the germination inducing activities of *P. ramosa* were significantly ($p < 0.01$) lower in tomato plants colonized by AM fungi than in free tomato plants (Lopez-Raez *et al.*, 2011).

When sunflower was inoculated simultaneously with *O. cumana* and AM fungi, a moderate level of protection against the parasite was observed. This protection did not only rely on a reduced production of germination stimulants. Mycorrhizal root exudates had negative effects on *O. cumana* germination (Louarn *et al.*, 2012). Root exudates from pea plants colonized by AM fungi showed lower germination inducing activities to *O. crenata*, *O. foetida*, *O. minor* and *P. aegyptiaca* than those from non-mycorrhizal plants (Fernández-Aparicio *et al.*, 2010b).
Pathogens play a certain role in the regulation of *Orobanche* under natural conditions. Some research work already focused on the use of these natural antagonists for the development of a mycoherbicide which could be a future tool in *Orobanche* control.

Novel approaches were recently developed to increase control by fungi, such as a "multiple-pathogen strategy." In this approach, two or more pathogens are used jointly and applied before or after parasite emergence. Some of these fungi mixtures resulted significant reduction in the number of emerged *O. cumana* shoots (Charudattan, 2001). Amsellem *et al.* (2001) and Cohen *et al.* (2002) reported a reduction in *O. aegyptiaca* attachment to tomato plants in greenhouse experiments using host-specific strains of *F.oxysporum* and *F.arthrosporioides*. The combination of Benzothiadiazole herbicide and *Fusarium oxysporum* f. sp. *orthoceras* successfully controlled *O. cumana* and reduced parasite emergence up to 100% (Müller-Stöver *et al.*, 2005).

Other techniques include the formulation or encapsulation of fungal propagules in a solid matrix to prevent rapid desiccation or microbial competition have been developed (Amsellem *et al.*, 1999; Quimby *et al.*, 1999). A successful example is Pesta granules that showed high efficacy in controlling *O. cumana* in the glasshouse (Kroschel *et al.*, 2000; Müller-Stöver, 2001; Elzein, 2003).

Another approach is engineering hypervirulence genes into weed-specific pathogens, *e.g.*, genes that encode enzymes degrade metabolites involved in parasite defense mechanisms such as phytoalexines, or coding for enhanced virulence by the production of fungal toxins (Gressel, 2002; Gressel *et al.*, 2004).

Chemical Methods

Different types and formulations of chemicals are used to control *Orobanche* species in different methods and times of application. These chemicals may be grouped into the following:

Fumigants

Volatile compounds such as methyl bromide have been used routinely to control *Orobanche* where the very high cost using this biocide could be justified. These fumigants are more confined to cash crop production such as in limited areas, under tunnels, plastic and glasshouses. Volatile compounds such as methyl bromide, ethylene dibromide, metham-sodium, or formalin are effective for controlling broomrape (Foy *et al.*, 1989), but the high cost seldom justifies their use in legumes or the open field and at a wide scale. Gayed (1986) reported on methyl bromide use also for *Cuscuta* in tobacco. Less expensive treatments that have been successful in recent years include formalin (1:50 formalin : water at 31 m^{-2}) under plastic for suppression of *O. aegyptiaca* in tomato in Iraq (Ella *et al.*, 1987), ethylene dibromide at 55-220 l ha^{-1} for control of *O. crenata* in pea (*Pisum sativum* L.), Metham-sodium at 10001 ha^{-1} for control of *O. aegyptiaca* in tomato (Jacobsohn *et al.*, 1987) which does not only kill *Orobanche* and other beneficial or harmful soil organisms including bacteria, fungi and nematodes but also harm personnel applying fumigants.

The practice of soil fumigation has caused environmental problems (Shabana *et al.*, 2003) and therefore, it was banned by the U.S. Environmental Protection Agency.

Herbicides

Generally, chemical control consists of treating the area with low rates of glyphosate (Roundup) and planting false host crops (Habimana *et al.*, 2014). Chemical control by systemic herbicides is also an option, although herbicides selectivity, in general, is a major issue.
Below are herbicide treatments recommended for parasite control in different crop species.

Seed Treatment

Seed treatments with imidazolinones proved effective for controlling *O. crenata* in faba bean. Coating sunflower seeds with 2 kg/ha and soaking the seeds in 50% of pronamide has lowered broomrape shoot dry weight and increased the yield of sunflower from 2.14 (control) to 2.85 t/ha in coated seed and from 1.24 (control) to 1.79 t/ha in soaked seeds (Sanchez *et al.,* 2003). Sulfonylureas also have the advantage of selectivity for preventing the emergence of broomrape growing on broad-leaved weeds in a non-host cereal crop: 3 g/ha metsulfuron-methyl, 15 g/ha chlorsulfuron or 22.5 g/ha triasulfuron gave 100% control of *O. ramosa* without damage to wheat or barley crops (Matthews, 2002). This may be due to their direct effect on *Orobanche* and to their reduction of broad-leaved weed hosts.

Soaking broad beans and lentils seeds for 5 minutes in 0.01% herbicide solution or coating at 20-40 g/ha (at a sowing rate of 160 Kg/ha) with imazethapyr did not affect seed germination and crop growth but resulted in 60-80% *O. crenata* control. Treatment using imazethapyr followed by late post-emergence application of imazapyr (5g/ha) resulted in more than 95% weed control. Effective seed treatments were found with imazapyr for broadbean and imazethapyr in lentils (Jurado-Exposito *et al.*, 1997). However, sulfonylurea, imazethapyr and imidazolinone were found effective in many host crops.

Soil and Foliar Applied Herbicides

Several reports are being published on the beneficial effect of mechanically incorporated herbicides belonging to dinitroanilines, sulfonylureas and substituted ureas groups, showing host crop selectivity and significant soil residuality for better control of *Orobanche* (Parker and Riches, 1993).

The imidazolinones herbicides are similar to the sulfonylurea in their general features and effectiveness. Different legume species resist these chemicals and, therefore can be selectively used to control weeds in these crops.

Split foliage application of different staff of imidazolinones herbicides proved effective and selective in controlling different *Orobanche* species including *O. ramosa, O. cumana,* and *O. crenata* in potato, sunflower and parsley, respectively. Three applications of imazapic through sprinkler irrigation using 4.5g/ha at 2 weeks intervals to potato root zone prevented *Orobanche* infestation and increased crop growth and yield, but tuber quality was severely affected in sandy soil (Goldwasser *et al.,* 2001). Split application of the same herbicide (2.5-5.0 g/ha) on parsley at 5-7 leaf before first cutting and on new growth after each cutting selectively controlled *O. crenata* and *O. aegyptiaca* (Goldwasser *et al.*, 2003).

Post-emergence of imazethapyr at 20 g/ha on pea (*Pisum sativum* and *Pisum arvense*) a month after planting and 20-40 g/ha two weeks later was effective against *Orobanche* and selective to pea plants (Jacobsohn *et al.,* 1998). Imazethapyr (75-100 g/ha) has been registered in Spain for use as a pre-emergence herbicide for *Orobanche* control in faba bean (Garcia-Torres *et al.,* 1998) and some promising results on the effectiveness of the same herbicide against *O. crenata* control through faba bean and pea seed treatments have been reported (Jurado-Exposito *et al.,* 1996; 1997; 1999). Rimsulfuron, a sulfonylurea herbicide, has also been registered as a selective herbicide for weed control in tomatoes and potato (Reinke *et al.* 1991). It is selectively controlled *O. aegyptiaca* in tomatoes when applied through drip irrigation to crop root zone. However, repeated application of this herbicide may be necessary for long-term weed control since it is of a short life in the soil (Kleifield *et al.,* 1994). Three foliar applications of rimsulfuron, each at 12.5 g/ha followed by irrigation, at two weeks intervals after crop emergence, were effective and selective in controlling *O. aegyptiaca* without damaging potato yield or quality (Goldwasser *et al.,* 2001). However, the use of rimsulfuron in tomato fields gave different results due to the differences in irrigation system used in both crops at which drip irrigation in tomato resulted in rapid leaching of the herbicide around the drip emitters. When applied at 10 to 20 g a.i./ha, rimsulfuron reduced broomrape dry weight and number of shoots and was selective on tomatoes (Vouzounis and Americanos, 1998) but toxic to eggplant. The same authors reported glyphosate and sulfosate applied twice at 30 to 50 g a.i./ha were effective against *O. ramosa* but reduced tomato yield. Some common pre-emergence herbicides, including pendimethalin, linuron, trifluralin, fluchloralin, metribuzin (PE/PPI), sulfosulfuron, oxyfluorfen, thiazopyr, isoproturon, chlorsulfuron and triasulfuron showed inconsistent effects against parasitic weed over the years and sometimes even showed phytotoxicity to the mustard crop or both (Yadav *et al.,* 2005). Seeds immersed with chlorsulfuron (0.05-0.1% solution) or triasulfuron (0.15-0.30% solution) for 5-10 minutes caused severe phytotoxicity on the emerged crop. In contrast, seed coating with lower doses (0.05-0.1 mg/kg) of chlorsulfuron, triasulfuron, or sulfosulfuron was safe for the crop. Mustard seed treatment with triasulfuron, sulfosulfuron and chlorsulfuron delayed *Orobanche* attachment and emergence, but the effect was inconsistent at longer periods. High doses of the herbicide seed treatment resulted in poor crop seed germination and growth suppression (Punia *et al.,* 2012).

<u>Broadbean</u>

Trifluralin applied pre-sowing and linuron post sowing gave the highest seed yield of broadbean (*Vicia faba*) infected with *O. crenata* (Zahran, 1982).

Americanos (1983) confirmed the effectiveness of glyphosate acid at 60-120g a.i./ha in controlling *O. crenata* in broadbean. Similar results were also obtained by Salem *et al.,* (1989), with glyphosate used at 52.5 and 178.5ml/ha and at 60g/ha at visible shoot buds or shoots and when roots were well developed at 130 and 452.4g a.i/ha (El-Hattab *et al*, 1982; Mesa-Garcia and Garcia- Torres, 1985).

Glyphosate applied at 0.12 kg/ha delayed parasite emergence and increased yield, but 2 applications were required in early sown crop and high parasite infestation (Mesa-Garcia and Garcia-Torres, 1982). Complete control of *O. crenata* was achieved with glyphosate used at 60 ml a.i./ha once or twice (13-16 days between treatments), and bean yield increased from 0-1.6 ton/ha and 0.3 to 2.3 ton/ha in plots treated with glyphosate (Schluter and Aber, 1980). When 2 sprays of glyphosate were applied at 64g/ha at the start of flowering and at the end of flowering seed and straw yields of cv. Giza 2 were increased by 1 and 1.18 t/ha, respectively (Nassib *et al.,* 1984), and very good results of *O. crenata* control were obtained when the herbicide was sprayed at 74g/ha at the start of flowering (Miccolis and Bianco, 1984). Nearly complete control of *Orobanche* was achieved with chlorsulfuron, imazaquine and glyphosate.

A similar study by Ibrahim *et al.*, (2000) revealed that glyphosate resulted in significant reduction in *Orobanche* spikes ranged from 92 to 100% compared with untreated plots. This high efficiency against *Orobanche* increased seed yield by 3,296 and 2,550 t/ha in two locations, respectively over the mean of all sites (Dongola *et al.*, 2000).

Results from studies of Hamid *et al.*, (2000) showed that the application of butralin at 2.85 Kg a.i./ ha followed by glyphosate at 58 g a.i./ha, imadazolinone at 214.2 g a.i./ha and hand hoeing (twice) were effective against *Orobanche*. These treatments reduced the fresh weight of *Orobanche* spikes by 95, 79 and 84%, respectively. Glyphosate proved effective against *O. cenata* in broadbean in Morrocco (Zerman, 2000), providing that the right time of glyphosate application has to be determined by inspections of the broadbean roots twice per week for *Orobanche* tubercules. The exact dosage of glyphosate is important for efficiency and to avoid phytotoxicity to the crop. Two treatments in an interval of 14 days have to be done. Even considering these points, the success of the application is not guaranteed. The effective dose to control *Orobanche* varies according to the numbers of tubercules attached and weather conditions. The use of glyphosate is well confirmed in resistant crops of faba bean, carrot, and celery.
Imidazolinones and sulfunylureas herbicides were also recommended for *O. crenata* control (Garcia-Torres and Lopez-Granados, 1991a). Nearly complete control of broomape in faba bean was achieved with chlorsulfuron, imazaquine, and glyphosate, as well. This high efficiency against broomrape resulted in a significant increase in faba bean seed yield by 3,296 and 2,550 t/ ha in two locations. A similar study on faba bean by Ibrahim *et al.* (2000) revealed that glyphosate significantly reduced *Orobanche* spikes by 92 to 100% compared to the untreated plots in all sites. For the late planting / low yield cropping systems, which normally face short and slight *crenata* infestations, crop pre- emergence infestations, imazethapyr at 75-100 g/ha and imazypyr at 12.5 g/ha resulted in effective control of the parasite without damaging broad bean. Imazethapyr, imazapyr and chlorsulfuron were the most effective for *O. crenata* control in legumes (Garcia-Torres and Lopez-Granados, 1991b), and imazethapyr was also effective against *O. aegyptiaca* and *O. foetida* (Geipert, 1997). Garcia Torres *et al.* (1998) reported selective *O.crenata* control in faba bean by pre-emergence and post-emergence applications of imazethapyr, imazapyr and imazaquin. According to Garcia-Torres and Lopez-Granados (1991b), early planting / high yield cropping systems which are normally subjected to heavy and lasting parasite infestations, would normally require a combination of pre-emergence treatments.

<u>Lentils (*Lens esculentus*)</u>
Glyphosate at 40 or 60 g/ha controlled *Orobanche* and increased yield of lentils by 117 and 51%, respectively (Arjona-Berral and Garcia-Torres, 1983). Effective seed treatments were obtained with imazethapyr (Jurado-Exposito *et al.*, 1997).

<u>Peas (*Pisum sativum*)</u>
Orobanche crenata caused yield loss in this crop by 100% at high infestation, and when *O. crenata* and *O. aegyptiaca* were stabilized it was at about 50% for moderate infestation (Bernhard *et al.*, 1998).

It has been previously reported that two or three applications of glyphosate at 150g/ha gave the highest yield under light infestation of *O. crenata* (Jacobsohn and Kelman, 1980). Ethylene dibromide alone or with chloropicrin controlled *O. crenata* and *O. cernua* (Jacobsohn *et al.*, 1982). while pre-emergence tank-mixture of propyzamide at 0.5 Kg/ha with methabenzthiazuron used at 2.5 Kg was also effective against *O. crenata* (Schlingloff and Alkamper, 1992).

Application of imazethapyr alone or imazethapyr with imazapyr at 5 g/ha to *Pisum sativum* seeds controlled *O. crenata* and seeds treated at late post-emergence resulted in excellent *Orobanche* control (Jurado-Exposito *et al.*, 1996).

Potato (*Solanum tuberosum*)

Single and sequential applications of rimsulfuron between 20 and 50 g a.i./ha significantly reduced *Orobanche ramosa* shoot number and dry weight compared to the control. Repeated applications (three times) were the most effective in reducing the parasite (Haidar *et al.*, 2005b). Best results considering both *Orobanche* control and selectivity in potato was achieved by application of rimsulfuron at 12.5 g a.i./ha followed by sequential foliar application (three times) of glyphosate at 100 g a.i./ha (Haidar *et al.*, 2005a). *Orobanche aegyptiaca* and *O. ramosa* were controlled in potato-infested soils by split foliar applications of low rates of the herbicides imazapic and rimsulfuron. Three doses of imazapic at 4.5 g/ha each, sprayed 2 weeks after crop emergence and re-applied at 2-week intervals, prevented *Orobanche* infestation (Goldwasser *et al.*, 2001). Chicken manure alone or with sulfur was effective in reducing *Orobanche ramosa* growth and infestation early in the season compared with the control. Unlike potato, the mixtures of chicken manure and sulfur at 8 and 12 t/ha significantly reduced *Orobanche* infestation later in the season (75 and 90 days after transplanting) in eggplant (Haidar and Sidahmed, 2006).

Vitch (*Vicia* spp.)

Nadal *et al.* (2008) reported that glyphosate used at 35–67 g a.i./ha, applied when *Orobanche crenata* attachment was at the tubercle stage of development, increased seed yield of narbon bean (*Vicia narbonensis*).

Cabbage (*Brassica oleracea var. capitata*)

Orobanche aegyptiaca was effectively controlled with glyphosate at 60-100 g a.i./ha or imazaquin at 5-10 g a.i./ha. The same authors indicated that trifluralin at 0.9 Kg a.i./ha may reduce *Orobanche* infestation (Americanos and Vouzounis, 1995).

Mustard (*Brassica juncea*)

Glyphosate applied twice using 25 g/ha at 30 days after sowing followed by 50 g/ha at 55 days after sowing provided 65-85% control of *Orobanche* in mustard (*Brassica juncea*) and up to harvest without any crop injury and improved yield from 12 to 41% over the traditional farmers' practice in different years of the study (Punia *et al.* 2010; Punia *et al.*, 2012).

Carrots (*Daucus carota*)

Glyphosate at 1-1.5 Kg/ha resulted in moderate control of *Orobanche* sp., and the herbicide was recommended for *O. crenata* and *O. aegyptiaca* control (Jacobsohn and Kelman, 1982).

Celery (*Apium graveolens*)

Infestation of both *O. ramosa* and *O. aegyptiaca* was reduced by spraying 20-50g/ha glyphosate. The most effective treatment was by using 50g glyphosate/ha (Americanos, 1991).

Parsley (*Petroselinum sativum*)

Orobanche crenata and *O. aegyptiaca* were completely controlled by split foliar application of imazapic at 2.5–5 gha^{-1} or glyphosate 36–72 gha^{-1}; applied on 5–7 leaf of parsley before the first cutting and on the young new growth after each cutting (Goldwasser *et al.*, 2003).

<u>Oilseed rape (*Brassica napus* L.)</u>
Foliar and soil applications of acibenzolar-s-methyl reduced *O. ramosa* attachment by 70% and prevented the loss of crop biomass (Veronesi *et al.*, 2009).

<u>Sunflower (*Helianthus annuus*)</u>
Soaking sunflower seeds in 0.1% boric acid or copper sulfate solutions increased the resistance of the plants to herbicides and to the attack by *Orobanche* sp. (Aziz, 1989). Treatment of sunflower seeds with 40 ppm of benzothiadiazole, Bion for 36 hours completely prevented the infection of *O. cumana* (Buschmann *et al.*, 2001). Propyzamide applied to sunflower plants in the greenhouse followed by sprinkler irrigation reduced or delayed *O. crenata* and *O. aegyptiaca* infestation. Adding the herbicide with irrigation water was the most effective method of control. Direct application of glyphosate and propazamide to the soil through an irrigation system was effective in controlling *O. cernua* (Kleifeld and Herzlinger, 1984).

Pre-sowing incorporation of propyzamide plus thiram into the soil prevented herbicide decomposition and kept the sunflower plants free of *Orobanche* until the end of flowering (Herzlinger and Kleifeld, 1985). Pre-emergence application of oxyflourfen at 1 and 2 l/ha reduced *O. cumana* density. Further reduction was obtained with pre-sowing treatment of trifluralin at 3.5 l/ha and pre-emergence application of oxyflourfen at 1l/ha. Oxyflourfen inhibited the formation of secondary haustorium of *O. cumana* (Horvath and Osztrogonac, 1991).

Imazethapyr, imazapyr, and chlorsulfuron were the most effective against *O. cernua* and thus imidazolinones and sulfunylureas were recommended for parasite control (Garcia-Torres and Lopez-Granados, 1991b). Iimazapyr at 10-15g/ha was effective in controlling *O. cernua* (Garcia-Torres *et al.*, 1995) and pre-emergence treatment of imazethapyr was found effective against the same parasitic species (Garcia-Torres *et al.*, 1998). In a post-emergence treatment, imazethapyr at 26.6-53.2 g/ha killed *Orobanche* nodules (Alonso *et al.*, 1998).

Imazapic reduced *O. cumana* infestation in post-emergence and sequential treatments in sunflower in both irrigated and nonirrigated fields while sunflower crop was not affected (Aly *et al.*, 2001).

<u>Tobacco (*Nicotiana tabacum*)</u>
Higher tobacco yield was obtained with 0.25 kg a.i. /ha maleic hydrazide (MH), which was in part with the hand weeding treatment in infested and not infested tobacco plants. Higher concentrations of MH, however, were toxic to tobacco crops. Foliage application of maleic hydrazide triethanolamine salt at 6 and 9 Kg/ha completely killed *Orobanche* spp. and the herbicide increased leaf yield and quality (Darbinyan *et al.*, 1977). MH reduced *Orobanche* spikes at 0.25 - 0.75 kg a.i./ha applied at 30 or 40 days after transplanting tobacco and up to the flowering stage (Papyan and Arutyunyan, 1981; Dhanapal, 1996).
Two applications of MH at 1-8 l/ha resulted in 90% control of *O. ramosa* in this crop (Danko, 1993a) while Imazapyr and EPTC were less effective. Imazaquin at 0.07+0.1 kg, maleic hydrazide at 0.45+0.45 Kg and sulfosate at 0.2+0.3 Kg, all reduced number of *O. ramosa* plants/pot compared with untreated control (Lolas, 1994).

Glyphosate and glyphosate trimesium (touchdown) at 100-150 ml/ha were also used to control *O. ramosa* in tobacco (Musselman, 1993). Glyphosate applied at 0.035% to the lower parts of tobacco plants in spot treatment was effective against *O. ramosa* (Danko, 1993b), and under field conditions, glyphosate in 2, 3 or 4 treatments with a total dose of 1000 ml/ha gave good control of

O. ramosa (Sandri *et al.*, 1998). At 0.5 kg a.i./ha applied at 60 days after transplanting and imazaquin at 0.01 kg a.i./ha applied at 30 days after transplanting reduced *Orobanche* population by almost 80% and increased tobacco leaf dry weight by more than 40% compared with the control (Dhanapal *et al.*, 1998).Tobacco gave a similar yield to weed-free treatment when 1.5 Kg fluchloralin, 0.75 Kg bentiocarb or 0.1 Kg oxyfluorfen and 4-inter-raw cultivation were done (Metha *et al.*, 1985). Pre-plant application of fluchloralin at 2Kg/ha gave a high yield of tobacco equal to the weed-free treatment. Fluchloralin and 4Kg diphenamid/ha were highly selective to tobacco, while pepulate and fluchloralin each at 3 Kg/ha were highly effective against *O. cernua* (Palled *et al.*, 1985).

Allyl alcohol at 0.1-0.2 concentration sprayed 4 times at weekly intervals reduced infestation of *Orobanche* (India, Central Tobacco Research Insitute, 1979). Chlorsulfuron (2g/ha), imazaquin (70 and 100g/ha), and imazapyr (20-30 g/ha) were applied pre-transplanting, and glyphosate (200-300g/ha), glyphosate-trimesium (300 and 400 g/ha), imazaquin (70 and 100 g/ha) and maleic hydrazide (450 g/ha) applied over tobacco plants were effective in controlling *Orobanche* with no phytotoxicity on the oriental cultivar (Lolas, 1997). The best control of *O. cumana* in tobacco was obtained with Treflan 48 EC (trifluralin 480 g l^{-1}) applied pre-emergence using 2 l ha^{-1} + Glyforom RV (glyphosate 360 g l^{-1}) post-emergence at 0.2 l ha^{-1} + 0.3 l ha^{-1} applied at 40 and 60 days from planting (Jinga *et al.*, 2005).

<u>Tomato (*Lycopersicon esculentum*)</u>

Glyphosate and sulfosate applied twice at 30 to 50 g a.i./ha were effective against *O. ramosa* and *O. aegyptiaca* in tomato and eggplant but reduced tomato yield. However, wide variations between tomato cultivars were detected for glyphosate treatment at 150 ppm. of glyphosate in 250l water /ha applied at 2-true leaf stage (Foy and Jacobsohn, 1983; Foy *et al.*, 1988).

Three applications (at bud formation, near flowering, and beginning of ovary formation) of glyphosate at 150 g/ha resulted in the highest tomato yield and lowest *O. ramosa* number (V"lchev *et al.*, 1995). Metham-sodium in drip-chemigation controlled *O. aegyptiaca* at 20-100 l/ha (Kleifeld *et al.*, 1991), and similar results were obtained by Goldwasser *et al.* (1993). Polyethylene mulching of the metham treated soil drastically improved metham performance and *O. aegyptiaca* control.

In pot experiments, chlorsulfuron applied at 37 and 180 g a.i./ha was the most effective treatment for *O. ramosa* control and the least toxic to tomato (Syka and Eleftherohorinos, 1991). *Orobanche aegyptiaca* control in tomato was achieved with 3-split applications of chlorsulfuron at 2.5 g a.i./ha and triasulfuron at 7.5 g a.i./ha (Hershenhorn *et al.*, 1998). Qasem (1998) reported excellent control of *O. ramosa* using chlorsulfuron at 2.44 g a.i./ha added with irrigation water or directly incorporated into the soil.

Bensulfuron, chlorsulfuron, nicosulfuron, primsulfuron, trimsulfuron, thifensulfuron and triasulfuron were tested for their effects on seed germination of *O. aegyptiaca* at the preconditioning and germination stages. All herbicides reduced radical elongation of the parasite. Chlorsulfuron and triasulfuron applied at 2.5 and 25 μM a.i. and rimsulfuron at 5μM a.i. at preconditioning and germination *in vitro* almost completely inhibited parasite development (Hershenhorn *et al.*, 1998). A single application of rimsulfuron (25 g a.i./ha) to tomato at 10 days after planting and split applications at 10 and 20 and 10, 20 and 30 days after planting significantly reduced the number of *O. aegyptiaca* inflorescences.

Chlorsulfuron and triasulfuron applied directly to the soil at rates $\geq$ 3.75 g a.i./ha completely controlled *O. aegyptiaca*. Possible contact phytotoxic effects on host foliage can be avoided using a very dilute solution and the herbicide washing off from foliage parts by sprinkler irrigation. Rimsulfuron at 10 to 20 g a.i./ha reduced *Orobanche* weight and number of shoots (Vouzounis

Americanos, 1998). Primisulfuron and chlorimuron, each at 22.5 g a.i./ha reduced the number of parasite inflorescence (Hershenhorn *et al.,* 1998).

Three split applications of 2.5 g/ha chlorsulfuron through sprinkler irrigation, starting two weeks after transplanting and at 10-14 days intervals, followed each application by 300 m^3/ha irrigation, controlled 80-90% of *Orobanche* with no phytotoxic effect on tomato. One or two split applications, through drip irrigation or at high volume were not sufficient to control the parasite during the whole growing season (Hershenhorn *et al.,* 1998c). To control late season *Orobanche* emergence around drip emitters in tomato, chlorsulfuron effectively controlled the parasite emergence (Kleifeld *et al.,* 1999), but results were inconsistent in other tests. Chlorsulfuron at 2.38g, 7.14g, and 11.9 g/ha gave 63.45%, 100% and 81.81% parasite control, respectively (Dongola *et al.*, 2000).

Sulfosulfuron was highly effective and selective for *O. aegyptiaca* control (Eizenberg *et al.,* 2001) but chlorsulfuron and triasulfuron were most effective through chemigation. Glasshouse experiments with activated charcoal suggested that the herbicide acts mainly through the soil and not by translocation through the host tomato plant. Successful chemical control of *O. aegyptiaca* in processing tomato was achieved with sulfosulfuron and imazapic (Parker, 2009). Two (14 and 42 days after planting) or three (14, 28, 42 days after planting) foliar applications of 50 g ha^{-1}effectively controlled *O. aegyptiaca* in tomatoes. These treatments were effective only for 70-80 days after planting. In India effective control of *Orobanche* in tomatoes grown under irrigation was achieved by pre-emergence application of sulfosulfuron at 75 g/ha (Dinesha *et al.,* 2012).

Deep incorporation of dazomet at 3 weeks before transplanting tomato prevented or delayed *Orobanche* emergence. The effect of dazomet, which releases the toxic gas methyl isothiocyanate to control *Phelipanthe mutelii,* was confirmed under field conditions in Australia (Prider and Williams, 2014).

Imazethapyr 20, 40 and 80 g/ha resulted in 77.72%, 72.7 % and 78% *Orobanche* control, respectively. Oxadiazon controlled *Orobanche* spp. when applied in sprinkler irrigation to the root zone of tomato (Kleifeld *et al.*, 1982). Imazaquin and glyphosate applied at 37 and 180 g a.i./ha, respectively, controlled *Orobanche* but imazaquin reduced crop yield (Syka and Eleftherohorinos, 1991).

Trials for chemical control of *O. ramosa* in tomatoes showed that imazethapyr and chlorsulfuron were effective for both crop and parasite situations (Dongola *et al.*, 2000). Application of the imidazolinone systemic herbicide (Cadre) on tomato foliage prevents fruit setting by damaging the reproductive system of tomato flowers throughout the growing season, with fruit set peaking at 50 to 60 days after planting. However, early treatments with monitor followed by Cadre application after the fruit setting peak completely controlled *O. aegyptiaca* in tomatoes without causing any damage to the yield (Lande *et al.*, 2005). Application of imazapic and imazapyr during tomato cultivation using an IR tomato mutant demonstrated high *P. aegyptiaca* control efficacy (Dor *et al.*, 2016). A foliar application of maleic hydrazide to tomato reduced *P. aegyptiaca* attachment on the host roots without any influence on tomato foliage or root dry weight (Samejima and Sugimoto, 2018).

Allelochemicals as Natural Herbicides for Orobanche Control

Orobanche Seed Germination Stimulants

Calcium hypochlorite and GA combined or used separately enhanced *O. crenata* seed germination (Pieterse, 1981). Immersion of 6- to 12-month-old seeds of *O. crenata* in 0.5-5.0 N H_2SO4 solution for 1-20 minutes increased germination of this parasite by 20% (Lopez-Granados and Garcia-Torres, 1996). Synthetic germination stimulants and strigol analogues (GR compounds) stimulated seed germination of different *Orobanche* species. Strigol has been reported for *Orobanche* stimulation (Vail *et al.*, 1985), and its analogues GR7 and GR45 stimulated germination of *O. minor*. It was found effective at 10^{-10} g/seed and GR45 was effective at 10^{-6} g/seed (Spelce and Musselman, 1981). The highest activity occurred at a concentration of 0.1-1 ppm. GR7, GR 24 and GR 28 at 1ppm were the most active of these compounds in inducing *O. ramosa* seed germination *in vitro*. Studies on the activity of various strigol analogues in different soils revealed that GR 7, GR 24 and GR 45 at 1ppm were most active. In a pot experiment, GR7 applied at 1-3 ppm. 6 weeks before transplanting tomatoes resulted in significant reduction in the dry weight and number of *O. ramosa* shoots and increased tomato yield (Saghir *et al.*, 1983). Under field conditions, GR7 at 0.3 Kg/ha resulted in good control of *O. crenata* in broadbean grown in acidic soils. The application of 1.5 Kg/ha of the stimulant controlled *O. crenata* in alkaline soils (Saghir, 1994). Strigol and Sorgolactone are also highly potent germination stimulants of *Orobanche* seeds (Willem *et al.*, 1997). Luque *et al.* (2000) reported that parthenolide and 3, 5-dihydroxydehydrocostus-lactone significantly increased *O. cumana* germination and presented higher activity than GR24.

Dhanapal (1996) reported that the germination phase of *O. cernua* in tobacco is a critical period. The seed bank of the parasite can be reduced by stimulating the germination through chemicals, natural host stimulants, or both. GR 24 at 1.0 and 0.1 ppm. was the standard to assess potential germination and reduction in the seed bank of *O. cernua* in tobacco. Among other chemicals, gibberellic acid at 10 and 20 ppm. was most effective. The stimulatory effects of host plants were significant even when GR24 was applied. The role of this hormone was further confirmed by Joel (1998), who reported that application of uniconazole as an inhibitor of gibberellin synthesis in pre-planting treatment to sunflower significantly reduced the number of infections by *O. cumana* and allowed the development of normal sunflower heads. Application of NAA and IBA induced an increase in root growth of *Solanum melongena* and enabled *O. cernua* to be maintained successfully. NAA at 60 ppm. gave best results (Rao *et al.*, 1982) and a recent study carried out by Harb (2002) confirmed the role that IAA in the maintenance of *Orobanche* on tomato roots. The activity of germination stimulants for suicidal germination of *Orobanche* seeds under field conditions could be substantially enhanced by applying brassinolide (2a, 3a, 22R, 23R-tetrahydroxy-24 S-methyl-B-homo-7-oxa-5-a-cholesten-6-ono) and related compounds to the infested soils (Takeuchi *et al.*, 1995). The herbicides fluridone (1-methyl-3-phenyl-5-(3-trifluoromethyl-phenyl)-4-(1-H)-pyridinone) and norflurazon (4-chloro-5-methylamino-2-(3-trifluoro methyl phenyl) pyridazin-3-one), inhibitors of carotenoid biosynthesis, were shown to induce GA-like effects on the conditioning and germination of *O. minor*. Accordingly, soil application of carotenoid biosynthesis inhibitors could potentially be a control method (Sang *et al.* 2004), by promoting seed conditioning of *O. minor* and other root parasites, and enhance the activity of germination stimulants to induce more effective suicidal germination.

It has been mentioned in general that strigol analogues have stability problems in the soil (Parker and Riches, 1993) and using them under field conditions may posses' different complications. In an experiment conducted by Georgieva and Bozoukov (1995) to test different pesticides for their ability to stimulate seed germination of different *Orobanche* spp., It was found that high stimulatory effects of the fungicides Topsin-M (thiophanate-methyl) (97%) at 50 ppm. and Ambis (90%) at 100 ppm., insecticides Thiodan (endosulfan) (80%) at 100 ppm. and Thiofanox (70%) at 10 ppm., biological pesticide Ditrapex (dichloropropane + methyl isothiocyanate) (90%) at 30 ppm., and the herbicide Vernam 6E (97.7%) (vernolate) at 200 ppm. and at higher than 1000 ppm. all showed inhibitory action to *Orobanche* seed germination.

Structure activity studies of the natural stimulants and their synthetic analogues that induce the induction of *Orobanche* seed germination proceeds via receptor-mediated mechanism (Zwanenburg and Reizelman, 2001) and the germination of root parasites depends on chemical products exuded from the roots of host plants known as germination stimulants. The predominant part of them belongs to sesquiterpenes lactones (Butler, 1995; Denev *et al.*, 2001). Yokota *et al.* (1998) isolated three seed germination stimulants of *Orobanche minor*, from the root exudates of its host *Trifolium pratense*. One was identified as alectrol and another was a novel strigol-related compound, which was named orobanchol and tentatively proposed to be a strigol isomer. Similar findings were reported by Yoneyama *et al.* (2001) in which at least three different germination stimulants to *O. minor* produced by red clover, including orobanchol and strigolactones, indicating that strigolactones are likely to be natural germination stimulants for *Orobanche* spp. Yoneyyama *et al.* (1998a) reported that jasmonates and related compounds elicited seed germination of *O. minor* and methyl jasmonate was the most active stimulant..

The identified natural germination stimulants of host root exudates with known chemical structures are strigol from maize, alectrol from cowpea, and sorgolactone from sorghum, and all produced by *Striga* host plants (Wegmann, 1998). Strigol is a powerful germination stimulant of *Striga* seeds (and later *Orobanche* seeds) that was first isolated and identified from the root exudates of cotton plants (a non-host crop) and stimulates seed germination of *Striga* by 85 and 100% at concentrations of 10^{-4} to10^{-6} M (Hsiao *et al.*, 1981). Other synthetic and natural germination stimulants of *Striga* species are lactonic or contain one or more unsaturated lactone rings (Worsham *et al.*, 1962; Johnson *et al.*, 1976; Musselman, 1980).

Some sesquiterpene lactones were tested by Perez-de-Luque *et al.* (2001). Some were found to induce *O. cumana* seed germination better than GR24 at low concentrations. They were specific in this parasitic species. However, some of these chemicals, such as coumarin-type compounds, are absolute allelochemicals (Worsham *et al.*, 1962; Rice, 1984). On the other hand, different species have shown strong ability to stimulate seed germination of different *Orobanche* species by more than 90% (Wegmann, 1998) but the structures of the germination stimulants exuded from these hosts remain unknown although several haustoria-inducing compounds have now been characterized including xenognosin A and B, identified in root exudates of *Astragalus gummifer*, and soyasapogenol B characterized from roots of *Lespedeza sericea*. More recent studies reported the sesquiterpene lactone group to play an essential role in stimulating seed germination of *Orobanche cumana* (Perez-de-Luque *et al.*, 2001).

Ethylene was reported by Zehhar *et al.* (2002) to be involved in germination induction of GR24 *Orobanche ramosa* conditioned seeds. Ethylene applied to conditioned seeds, strongly reduced induction of germination by GR24. However, exogenous ethylene did not induce germination of conditioned seeds, but 2-chloroethylphosphonic acid was able to do so. When inhibitors of gibberellins biosynthesis were applied to conditioned seeds in the presence of GR24, parasite germination was inhibited. The same chemicals also strongly inhibited germination of conditioned *Striga hermonthica* seeds in response to GR24 but this inhibitory effect was reversed by the addition of 2-chloroethylphosphonic acid. The effect of these inhibitors on *S. hermonthica*, in which ethylene is a necessary mediator of germination induction by GR24, strongly suggests that ethylene synthesis is also required for the induction of *O. ramosa* seed germination by GR24. These growth regulators, which inhibit the two steps of germination in *O. ramosa*, could be useful for the development of methods for early season control of this parasite.

Willem *et al.* (1997) synthesized two diastereomers of demethylsorgolactone. The germination stimulatory activity of these compounds is comparable to that of the strigol and that there exist significant differences in activity among the individual stereoisomers. A newly designed strigol analogue Nijmegen 1 was prepared to start from N-phthaloylglycine (Nefkens *et al.*, 1997) and was found active at low concentration as a suicidal germination agent for both *Striga hermonthica* and *Orobanche crenata* (Wigchert *et al.*, 1999). Nijmegen 1 was resolved into its rac and ent enantiomers, which were found to exhibit significant differences in germination activity. It is evident now that allelochemicals play an important role in increasing or decreasing parasitic weed problems and can be utilized as promising and effective natural tools in controlling these species. Some of these allelochemicals are seed germination inhibitors that can be used in a pre-planting soil applied treatment to inhibit parasite germination or attachment to host plants, while others are able to stimulate parasite seed germination and thus can be used in a similar way but in the absence of host species. However, the amount, concentration and environmental conditions are significant factors that determine the effectiveness of these chemicals. Sugimoto (2000) reported different natural stimulants for different parasitic species, including strigol from cotton, sorgolactone from sorghum, alectrol from cowpeas, and red clover (*Trifolium pratense* L.) root exudates (Xie *et al.*, 2008), and orobanchol from red clover (*Trifolium pratense*). Chittapur *et al.* (2001) discussed and evaluated the use of catch and trap crops in integrated weed management and emphasized their role in controlling different parasitic species of economic importance through allelopathy mechanisms. Orobanchol and alectrol are natural germination stimulants of *O. minor* that are produced by red clover (Yoneyama *et al.*, 2001); these indicate that strigolactones are likely to be natural germination stimulants for *Orobanche* spp. Several classes of natural germination stimulants were isolated from root exudates of host and non-host plant species of *Orobanche*. Some identified as strigolactones (Xie *et al.*, 2010), peagol and peagoldione (Evidente *et al.*, 2009), peapolyphenols A–C (Evidente *et al.*, 2010), soyasapogenol B, trans-22-dehydrocampesterol (Evidente *et al.*, 2011), dehydrocostuslactone (Joel *et al.*, 2011), or isothyocyanates (Auger *et al.*, 2012). Strigolactones have been identified as germination stimulants for root parasites (Yoneyama *et al.*, 2009) and play a major role in host specificity (Fernández-Aparicio *et al.*, 2011). Orobanchol, orobanchyl acetate, and 5-deoxystrigol are widely distributed in the Fabaceae (Yoneyama *et al.*, 2008). Fabacyl acetate has been identified from root exudates of pea (Xie *et al.*, 2009). Peagol and peagoldione have also been detected in pea root exudates and found to selectively stimulate *Orobanche* seed germination (Evidente *et al.*, 2009). The same root exudates were found containing peapolyphenols A, B and C, in addition to a polyphenol and a chalcone. Only peapolyphenol A, 1,3,3-substituted propanone and 1,3-disubstituted propenone had specific stimulatory activity on *O. foetida* but not on other *Orobanche* or *Phelipanche* tested species (Evidente *et al.*, 2010). Root exudates of common vetch were found containing a triterpene and a sterol, identified as syasapogenol B and trans-22-dehydrocampesterol. Soyasapogenol B was very specific in stimulating germination of *O. minor* seeds only, while trans-22-dehydrocampesterol stimulated seed germination of *P. aegyptiaca, O. crenata, O. foetida,* and *O. minor* (Evidente *et al.*, 2011). Root exudates of oilseed rape contain 2-phenylethyl isothiocyanate that was found to stimulate seed germination of *P. ramosa* (Auger *et al.*, 2012). Root exudates of black oat include avenaol, a potent germination stimulant of *P. ramosa* seeds but have a weak activity to *O. minor* (Kim *et al.*, 2014). The structure of a major strigolactones, produced by *Medicago truncatula* was determined and named medicaol (Tokunaga *et al.*, 2015).

The role of microorganisms (*e.g. Streptomyces*) has also been implicated in *Orobanche* seed germination. Christeva and Naumova (1998) reported that different strains of *Streptomyces* induced germination of *O. ramosa* and concluded that microorganisms might take part in the germination processes of this parasite. In another study, Yoneyama *et al.* (1998b) reported stimulation of *O. minor* seed germination by certain fungal metabolites including cotylenins and fusicoccins at concentrations as low as 10^{-5} M and up to more than 50%. Different *Fusarium* spp. were reported by Sauerborn (2001) to infect *Orobanche* and *Striga* seeds and plants, and the author emphasized the role of phytotoxins in the process. He added that members of the genus *Fusarium* produce toxins such as enniatin, fumonisin, fusaric acid, moniliformin, and trichothecenes that possess a broad range of biological activities and metabolic effects. However, some of these compounds have already been considered and proposed to be used as natural herbicides (Strobel *et al.*, 1991).

Khalaf *et al.* (1997) studied *Orobanche crenata* control using *Azotobacter* spp. and *Escherichia coli* transformants that break parasite seed dormancy. Using crude DNA preparations from *V. faba* roots to prepare transformants of *Azotobacter* spp. and *E. coli,* the transformants produced *Vicia faba*-like stimulants, which promoted seed germination of the parasite. Transformants of *A. chroococcum* and *E. coli* were most active. The greatest levels of *O. crenata* germination achieved in the bioassays with transformed colonies of *A. chroococcum* and *E. coli* were 41.8 and 27.3%, respectively. The authors proposed that such transformants have potential as biological control agents by causing non-host germination of the parasitic weed seeds. Fusicoccin and cotylenol proved to induce seed germination of *O. minor.* Evidente *et al.* (2006) tested 25 natural analogues and derivatives of fusicoccin for the ability to stimulate seed germination of *O. ramosa* and found some of the compounds tested proved to be highly active, being 8,9-isopropylidene of the corresponding fusicoccin derivatives aglycone and the dideacetyl derivative were the most active. Fusicoccin is a toxic metabolite produced by the fungus *Fusicoccum amygdali*, the causative fungal agent of peach and almond canker (Boari *et al.*, 2005). Nijmegen^{-1} has been tested as a suicidal germination agent in field trials on tobacco infested by *O. ramosa* and results were highly promising (Zwanenburg *et al.*, 2009; Zwanenburg and Pospíšil, 2013).

Natural Orobanche Seed Germination Inhibitors

Inhibitory allelochemicals may work at different stages of the parasite life cycle from germination to growth and development. However, a reverse action may be obtained at a low concentration received by the parasite. Sauerborn (2001) suggested that phytotoxins could contribute to improving parasitic weed management. Serghini *et al.* (2001) reported differences between *O. cernua*-resistant and susceptible sunflower varieties with seed germination of *O. cernua* halved and germinated seeds displayed enhanced browning symptoms in the presence of resistant sunflower roots. Parasite radicles or host-tissue around the contact point turned brown after *O. cernua* attachment, especially to roots of the resistant varieties. They suggested the possible accumulation of toxic compounds as a defense strategy in the resistant sunflower varieties. Serghini *et al.* (2001) stated that coumarins affect the normal growth and development of *O. cumana* seedlings. The effect was more intense from the resistant cultivars than in the susceptible ones. Coumarins (ayapin and scopoletin) from sunflower plants were implicated as inhibitory allelochemicals to germination, growth and development of *O. cumana* (Perez-de-Luque *et al.*, 2001) and both are released by sunflower roots into the environment and could act as toxic allelochemicals to the parasite (Guitierrez-Mellado *et al.*, 1995). Sunflower 7-hydroxylated simple coumarins may play a defensive role against *O. cernua* parasitism by preventing successful germination, penetration and/or connection to the host vascular system. This hypothesis is supported by that coumarins inhibited the *in vitro* germination of *O. cernua* seeds induced by the strigol analogue GR24 and caused a browning reaction in germinated seeds. Resistant sunflowers accumulated higher levels of coumarins in roots and excreted greater amounts than susceptible varieties in response to *O. cernua* infection. Bar-Nun and Mayer (2002) reported the presence of large amounts of oleic and linoleic acids and small amounts of stearic and palmitic acids in *O. aegyptiaca* seeds. The authors expected these acids to interfere with sugar metabolism and thus to prevent parasite seed germination during the preconditioning period. Pre-plant treatment of crop seeds with these inhibitors may prevent parasite infestation and could be considered as an effective technique in parasitic weed control. Sunflower seeds treated with 40 ppm. of benzo (1,2,3) thiadiazole-7-carbothioic acid S-methyl ester (BTH) for 36 h completely prevented infection of *Orobanche cumana* in root chambers (Sauerborn *et al.,* 2002). In pot studies, at considerable inoculums of *O. cumana* seeds, the total number of parasite shoots was reduced by almost 90% with 60-ppm BTH. The inhibitory chemical prevented both attachment and penetration of *O. cumana* at the sunflower root. Results showed the synthesis of the phytoalexin scopoletin and of hydrogen peroxide in the BTH-treated sunflower roots and demonstrated the importance of allelopathy strategy as an effective component of future plant production systems.

The inhibitory role of microbial phytotoxins has been very well documented. Zonno and Vurro (1999) found that T-2 toxin from *Fusarium* sp. to be the most active among 14 fungal toxins tested, inhibited by 100% seed germination at 10^{-5} M and was active down to 10^{-7} M (19% inhibition). Amsellem *et al.* (2001) mentioned that mycotoxins isolated from different *Fusarium* and other fungal species killed *Orobanche*, and are being considered for direct use or to complement other control strategies. *Fusarium* spp. have been transformed with gus and/or gfp genes allowing tracing their movement in the environment and opening the way to future transformations to hypervirulence.

In another study, Zonno and Vurro (2002) tested the effect of toxins produced by *Fusarium* species on *O. ramosa* seed germination and found many were active at the highest concentration. Fusarenon-X, nivalenol, deoxynivalenol, T-2 toxin, HT-2 toxin, diacetoxyscirpenol and neosolaniol were highly active at 10 μM, causing 100% inhibition of seed germination. Many of the toxins tested were active when assayed at a concentration ten times lower, with T-2, HT-2, nivalenol, neosolaniol and diacetoxyscirpenol still able to cause complete damage inhibition. Diacetoxyscirpenol was also very active at 0.1 μM, causing more than 90% inhibition. Results showed that the use of toxic secondary metabolites could represent a useful alternative strategy in the management of parasitic weeds by interfering with the induced germination process, and that fungal culture extracts could be an interesting source of new compounds acting as natural and original herbicides on these parasites.

Chen *et al.* (2016) reported that *Streptomyces enissocaesilis* significantly reduced the germination rate of *O. cumana* seeds both in the seed germination experiment and in the co-culture experiment, with more than 50 and 40 % (after cultivation for eight days) reduction, respectively over the control. In the pot experiment, application of *S. enissocaesilis* reduced the epigeous number of *O. cumana* tubercles by 47.5 % after 130 days. Furthermore, *S. enissocaesilis* also increased polyphenol oxidase activity in sunflower roots and improved the beneficial microflora in the rhizosphere soil of sunflower.

Ancymidol, uniconazole and paclobutrazol were reported as strong inhibitors of *O. ramosa* seed germination and CCC, daminozide and prohexadione at 10^{-5} had a similar effect (Zehar and Fer, 2001). The inhibitory effect of paclobutrazol and uniconazole could have resulted from the increased level of abscisic acid (ABA). Oxidative metabolism of ABA into phaseic acid and exogenous ABA is a strong inhibitor of *O. ramosa* germination (Pouvreau *et al*., 2013).

Other Allelopathy Forms and Application Methods

Plant Extract

The effect of different plant extracts in stimulating or inhibiting *Orobanche* seed germination or parasite attachment was investigated by different workers. Lopez-Granados and Garcia-Torres (1996) studied the effect of root extract stimulants of broad bean plants on *Orobanche crenata* seed dormancy and germination and found germination stimulant activity during different seasons were varied considerably and decreasing from autumn to summer seasons, suggesting an ecological adaptation of *O. crenata* to its host crop.

Egyptian clover was reported as an effective species in stimulating seed germination of *O. crenata*. Extract of this crop induced parasite germination by 98% followed by chickpea extract (95%) and lupine extract (91%), and seed germination was found better in sandy soils than in clay soils (Zaki *et al.*, 1998). Eizenberg *et al.* (2001) reported root extracts of sunflower resistant cultivars to inhibit the development of *O. cumana*, while methalonic crude extract exhibited the strongest phytotoxic activity. Results indicated the possible role of phytoalexines in the resistant reaction. Ma *et al.* (2012) tested 606 extracts of Chinese medicinal herbs to stimulate germination of *Orobanche minor*, *Orobanche cumana*, and *Orobanche aegyptiaca*. Results showed that *O. minor* of a narrower host spectrum exhibited a wide range of germination responses to the various extracts. *Orobanche cumana* and *O. aegyptiaca* showed an intermediate germination response. Methanolic extracts of many herbal species effectively stimulated seed germination among the *Orobanche* spp., although they were not the typical hosts. The effective herbs represent interesting examples of potential trap crops. Dong *et al.* (2012) reported that methanolic extracts of all tested wheat varieties displayed a stronger induction of *O. minor* seeds germination than the aqueous extracts. The aqueous extracts at 10% and 1% effectively stimulated parasite seed germination, but full strength (100%) and 0.1% extracts did not induce any seed germination. The methanolic extract of variety `Xiaoyan No. 22' induced *O. minor* seed germination up to 56.8%. Thus *O. minor* seeds could be induced to germinate by the aqueous and methanolic extracts of winter wheat varieties due to their contents of allelopathic substances at the seedling and heading stages. The efficacy of the natural product, extracted from the alga *Ascoplyllun nodosum* (Anextraction), was evaluated for the induction of *O. ramosa* seeds germination and found of possible potential use as a germination stimulant for broomrape control (Lyra *et al.*, 2005)

Plant Oils

Application of plant oils (gingelly, groundnuts, palm, sunflower, safflower, niger, castor, linseed, neem, coconut or tobacco seed oils) at a rate of two or three drops to the top of *O. cernua* killed parasite shoots in 2-4 days (Krishnamurthy, 1992). However, oil treatment needs to be repeated every 4-5 days and for 3-4 times on emerging shoots. Botanical pesticides from neem (*Azadirachta indica*), *Pongamia*, *Nicotiana gossei* and nicotine sulfate were also recommended for post-emergence control of *O. cernua* with plant oils (Chari *et. al.*, 1993). Putting 1-2 drops of diesel oil, boiling water and kerosene oil on each shoot have also been suggested for the control of this parasite (Krishnamurthy *et al.*, 1976; Linke and Saxena, 1991b). Similarly, oils of gingelly, groundnut, palm, sunflower, safflower, niger, castor, linseed, coconut, tobacco, eucalyptus, pongamia, soybean, rice bran *etc.* applied 2-3 drops on the top of heads have been reported to kill broomrape shoots within 2-4 days, but with less effect on flowering shoots (Krishnamurthy and Chari, 1991; Krishnamurthy and Nagarajan, 1991b; Krishnamurthy, 1992). Not all oils were quite effective, but they have the advantage of not being phytotoxic to the host plant. However, these techniques have practical problems and 3-4 repeated applications on emerging shoots at an interval of 4-5 days are required for its effective control. All these aforesaid oils cause only localized desiccation and prevent seed setting but later on emergence of other shoots was observed. Swabbing natural plant oils killed the bud and stem parts of the parasite by suffocation. The effect was selective on different parasite parts. While neem, coconut and sunflower oils showed quick knockdown effects in killing the parasite bud part., neem oil did not kill the stem part of the parasite, and niger (*Guizotia abyssiniaca*), castorbean and mustard oils appeared to be somewhat less effective (Dhanapaland Struik, 1998). In another study, *Azadirachta indica*, coconut and sunflower oils affected the buds of *O. cernua* within 3 days; *Ricinus communis* and *G. abyssiniaca* oils killed buds in 3-4 days; and mustard oil took 5 days to kill buds. Coconut and sunflower oils killed *Orobanche* stems more rapidly than niger and castorbean oils with no observed phytotoxicity on tobacco (Dhanapal *et al.*, 1998). Volatile oils and crude acetone extracts of different plant species were also effective against *O. ramosa* under glasshouse conditions (Solymosi, 1998).

The Use of Advanced Biotechnology

Nanotechnology

In the near future, some nanotechnologies can improve existing crop management techniques. Nanocapsules would help to avoid phytotoxicity on the crop by using systemic herbicides against parasitic weeds. Nanoencapsulation can also improve herbicide application, providing better spread and penetration through cuticles and tissues, and allowing the slow and constant release of the active substances.

Nanoparticles have great potential as 'magic bullets', loaded with herbicides, chemicals, or nucleic acids, and targeting specific plant tissues or areas to release their charge. Viral capsids can be altered by mutagenesis to achieve different configurations and deliver specific nucleic acids, enzymes or antimicrobial peptides acting against the parasites. Many issues are still to be solved, such as increasing the scale of production processes and lowering costs, as well as toxicological issues, but the foundations of a new plant treatment concept have been laid, and applications in the field of parasitic plant control may be started (Pérez-de-Luque and Rubiales, 2009).

Genetic Engineering

Genetic engineering technology enables the use of herbicides, including the non-selective chemicals on crop plants and targeted weed species, providing that the herbicide-resistant gene is transferred into crop plants through this technology (Qasem, 2013). Examples below present different engineered crop species to resist herbicide treatments and to target *Orobanche* species.
Transgenic resistance to herbicide basta (glufosinate) could be used as a post-emergence agent for *Orobanche* spp. control (Valkov *et al.*, 1998). Tomato plants engineered with the bar gene conferring metabolic resistance to glufosinate, an inhibitor of glutamate synthase, tobacco plants containing the SuRB-Hra gene encoding a modified acetolactate synthase (ALS) gene conferring target-site resistance to chlorsulfuron. Rape plants containing the aro-A gene encoding a modified enolphosphate-shikimate phosphate synthase conferring target-site resistance to glyphosate and tobacco plants containing the sul-1 gene conferring target-site resistance to assulam (Joel *et al.*, 1995). *Orobanche* control in transgenic tobacco cultivars Elenski 817, Melnik 812, Djebel and Jaka resistant to glufosinate ammonium was achieved at 2-20 l/ha and with chlorsulfuron 60-180 g/ha (Batchvarova *et al.*, 1998). Promising results on tobacco resistance to *O. ramosa* were also reported after treating tobacco seeds with N-Nitrosoethylurea or Ethylmethanesulphonate solutions (Slavov *et al.*, 2001).

Glyphosate proved excellent in suppressing *O. aegyptiaca* in both genetically engineered *Vicia sativa* and on EPSPS inhibitor-resistant oilseed rape (*Brassica napus*) (Nandula *et al.*, 1999). Transgenic assulam – potatoes with the herbicide resistance gene sul that codes for a modified dihydropteroate synthase (DHPS) as a method of weed eradication was recommended. *Orobanche* was completely suppressed on the best potato transformant when assulam was used at different concentrations (0–700 mM) and at different times after shoot emergence and potato viability were preserved (Survo *et al.*, 1998). In transgenic potato plants, *O. aegyptiaca* was completely controlled by assulam without damaging potato plants (Joel *et al.*, 1999).

Aviv *et al.* (2002) transferred a mutant ALS gene into the carrot, which allowed the control of broomrape by imazapyr. Aly *et al.* (1999) reported preliminary results on germination and radical elongation inhibition of *O. aegyptiaca* seeds by sarcotoxin IA peptide obtained from yeast. In a later study, Aly *et al.* (2006) reported enhanced resistance to *O. aegyptiaca* when the sarcotoxin IA gene was linked to the constitutive root-specific Tob promoter to generate sarcotoxin-expressing tomato plants. The attached parasite showed abnormal growth and high mortality while tomato plants were growing normally. Sarcotoxin however, is an antimicrobial polypeptide produced by the fly *Sarcophaga peregrina* (Kanai and Natori, 1989). The transgenic herbicide resistance may impose food safety problems since done by transferring a strange gene to a crop plant. Gene transfer from transgenic crop plants to other wild-grown plant species and parasitic species, including *Orobanche* is possible. Parasitic weeds could rapidly develop resistance to herbicides because of their huge seed production. The use of herbicide resistant crops must be very well managed and integrated with other control methods, and new resistant crops are continually developed.

Transgenic allelopathic crops that could prevent parasite seed germination, reduce or delay the release of parasite seed germination stimulants, or prevent attachment of the parasite or haustorial development. This would require gene identification mainly in trap or nonhost crop species and an extensive search on their possible insertion in crop plants. Genetic engineering of inhibitory and stimulatory chemicals, as well as genes control of other resistance/tolerance mechanisms involved, could be also investigated. In summary, herbicide resistant crops were developed in tobacco for different *Orobanche* species control (Acetolactate synthase (ALS and Chlorsulfuron)), carrot for *O. aegyptiaca* control (Imazapyr), tobacco and potato for *Orobanche* spp. using Asulam herbicide (Aly, 2007).

Gene Silencing

This is a new advanced technology (Thijn *et al.*, 2002 ; Galun, 2005) effective through a segment of a double-stranded RNA (dsRNA) inserted in host plants and can move into the parasite and inhibit parasite target gene leading to its impaired expression (Fire *et al.*,1998), preventing its function and reducing parasitism. Dubey *et al.* (2017) suggested this strategy for controlling parasitic weeds based on the simultaneous trans-specific gene-silencing of three parasite genes. Two strategies were used to express dsRNA containing selected sequences of *P. aegyptiaca* genes PaACS, PaM6PR, and PaPrx1 (Rubialis *et al.*, 2018). The use of RNA interference was shown as of a high potential strategy for controlling plant parasitism (Tomilov *et al.*, 2008). Host–induced gene silencing has been used against different parasitic species such as *Triphysaria versicolor* and *Cuscuta pentagona* and proved effective in controlling their infestation to host crops (Alakonya *et al.*, 2012; Bandaranayake and Yoder, 2013; Yoshida *et al.*, 2016). However, crop plants treated through this and other biotechnology methods such as genomic, proteomic, and metabolomic must be safe for humans and the ecosystem. The engineered herbicide resistance crops may have food safety problems. A concern may also arise regarding the possible gene transfer to weeds through gene exchange in the field. Gene silencing may provide crop plants with defense against parasitic weeds, including *Orobanche,* since it acts against intervening plants pathogens and is an important tool for research on plant development (Hunter and Poethig, 2003). This strategy is best since effective, cheap in the application, and eco-friendly.

It has been shown that the reciprocal movement of silencing RNA could be from parasite to host. Targeting genes in *P. aegyptiaca* resulted in the impairment of parasite growth on the engineered hosts (Dubey *et al.*, 2017). As for herbicide efficacy, the reason for the differences among parasites in response to RNAi expression in host plants could be due to the nature of the vascular connections formed in the different parasite interactions. Infection by stem parasites can also be inhibited using cross-species RNAi. RNA interference transgenic tobacco results in defects in the establishment of *Cuscuta* haustorium (Alakonya *et al.*, 2012). Until now, only a few reports are available on gene editing uses for possible control of parasitic weeds. Those that have appeared targeted pre-attachment resistance. Recently, Bari *et al.* (2019) reported that CRISPR/Cas9-mediated mutagenesis of the CCD8 gene in tomatoes resulted from low root stimulation activity of parasitic seed germination because of the loss of root strigolactone, the main crop germination stimulant, and a significant reduction in parasite infestation compared to wild type tomato plants. However, some negative agronomic effects of the gene-editing can be resulted, including dwarfing, excessive shoot branching, and adventitious roots.

Integrated Control Methods

The only effective way to counteract parasitic weeds problems is to apply an integrated approach (Rubiales and Fernández-Aparicio, 2012). Integrated control of parasitic weeds means a combination of all possible weed control methods and tools. These are the preventive methods that not allow parasite introduction into the field and by all means; cultural methods of *Orobanche* control that help avoid germination and improve crop tolerance (Sowing date, ploughing, altering planting dates, crop rotation, trap and catch crops, resistant varieties, screening for field tolerance and identification of trap crops, and fertilization and varying fertilizer application); mechanical (Hand weeding, tillage, cutting); Physical (soil solarization, mulching and soil covers, flooding, irrigation, burning); biological (Insect and pathogens); and future research (new herbicide treatments, pathogens, biotechnologies, transgenic plants, germination inhibitors and allelochemicals and herbicides applied with crop seeds); and chemical methods of control (Glyphosate, pre-emergence herbicides, new post-emergence herbicides and soil fumigants/sterilants, germination stimulants) (Garcia-Torres, 1993; Krishnamurthy, 1994; Dhanapal *et al.*, 1996; Fer and Thalouarn, 1997) .

Not all satisfactory means of control have been found so far; therefore a combination of suitable measures for a given situation in the sense of an integrated approach seems to be the most appropriate way to deal with the problem (Dominguez *et al.*, 1993). The results obtained from any single control method were insufficient. Combining two or more methods can tackle the problem easier. In this regard, analysis of the farming system looks important (http://www.uni-hohenheim.de/~www380/parasite/oro_path.htm).

At present, control procedures consist of a combination of mechanical (hand removal), chemical (glyphosate), and cultural procedures (planting false host crops) (Habimana *et al.*, 2014). For *O. cumana* control program in sunflower, resistant cultivars, deep ploughing and seed cleaning have been suggested (Petzoldt *et al.*, 1993). Early sowing (mid-October) and an annual application of 75g/ha imazethapyr and 40g/ha glyphosate was the best method for weed control in broadbean (Lopez-Granados *et al.*, 1997). Glyphosate with Nuflim-P (14.4g + 40cm^3) resulted in the greatest seed yield of broadbean and glyphosate, plus Greenzit-NPK (21.6g + 60cm^3) gave the greatest straw yield (Nassib *et al.*, 1990). Glyphosate 34g a.i./ha with NPK fertilizer effectively controlled *O. crenata* (Hussein *et al.*, 1998).

In tobacco, hand weeding, crop rotation, trap crops, catch crops with solarization, *P. orobanchia*, resistant varieties, fumigants, stimulants (ethylene) and herbicides (maleic hydrazide, glyphosate, sulfosate, imazaquin and imazapyr), all were recommended for integrated control of *O. ramosa* and *O. cernua* (Lolas, 1998). Cultural and chemical integration methods for *Orobanche* sp. control in lentil, chickpea, faba bean, Narbon vetch, vetch, forage legumes and field pea were proposed by ICARDA (1986).

Combined treatment of the herbicide benzothiadiazole with the pathogen *F. oxysporum* f. sp. *orthoceras* successfully controlled *O. cumana* and reduced parasite emergence up to 100% (Müller-Stöver *et al.*, 2005).

The highest level of protection against the parasitic weed was achieved on Hashenge variety of faba bean, which also meant that the same variety had the best agronomic performance and the highest yield so that it may be used in combination with *Trichoderma harzanium* and *Rhizobium leguminosarum* as an integrated management option against *O. crenata* (Brhane *et al.*, 2016; Aregawi, 2017).

BROOMRAPES (*OROBANCHE* SPP.) IN JORDAN

GENERAL BACKGROUND

Two genera of the Orobanchaceae family occur in Jordan flora (Qasem, 2009); these are *Orobanche* and *Cistanche* and species of both are root parasites. *Orobanche* species are widely spread in different parts of the country, attacking a large number of host species of different growth habits and habitats, including field crops, vegetables and fruit and forestry species. *Cistanche* species have different morphological features that distinguish them from *Orobanche.* Typically, they parasitize higher life forms, mostly including trees and shrubs. Both are more of warmer regions of the world. *Cistanche* is more adapted to desert tropics, subtropics, and saline areas while *Orobanche* spread under a wider climatic condition.

It has been previously stated that 11 *Orobanche* species were recorded in Jordan; these were presented in Table 1. However, in a more recent work, Qasem (2009) reported the presence of seven species of *Orobanche* in cultivated land in the country attacking 86 plant species belonging to 24 plant families in different regions of Jordan (Table 1.1.5) and their distribution between different species of the parasite is shown in Table 1.1.6. Most parasitized families: Compositae (20 species), Solanaceae (11 species), Leguminosae (9 species), Umbelliferae (7 species), Cruciferae (7 species), Cucurbitaceae (4 species), Labiatae (4 species) and Rosaceae (4 species). Among those parasitized, 50 were crop species. Many of the host species identified are first time known. *Orobanche ramosa* was the most common parasite with the widest host range (73 species). Screening studies of different summer and winter crops, medicinal herbs, and a large number of weeds (Qasem and Foy, 2007) for a possible attack by this parasite, revealed substantial variation among species in the number and vigour of parasites attached to their roots (Table 1.1.5). The crops and medicinal herbs most affected were: *Apium graveolens*, *Carthamis tinctorius*, *Carum carvi*, and *Trachyspermum ammi* (with 36–214 *Orobanche* shoots per pot). Among the crops studied, *Ammi visnaga*, *Brassica nigra*, *B. oleracea* var. *botrytis*, *Cucurbita moschata*, *Linum usitatissimum*, *Lupinus albus*, and *Raphanus sativus* were least infected (with 1– 2 *Orobanche* shoots per pot).Weeds, including *Datura metel*, *Ferula communis*, and *Rumex acetosella* were heavily infected (with 40–67 shoots per pot). In contrast, *Antirrhinum orontium*, *Convolvulus arvensis*, *Diplotaxis erucoides*, *Polygonum aviculare* and *Solanum cornutum* showed extremely low infection (with 1–2 shoots per pot). Other plants tested may be useful as “trap or catch species” and both may be used effectively to manage infestation by the parasite by their inclusion in crop rotation or intercropping systems. Non-host species or those that allowed very low parasite attachment might be grown in fields heavily infested by the parasite. Control of highly infected weed species is necessary to prevent parasite multiplication and spread. However, timing this operation after parasite attachment or before emergence may be effective in exhausting the soil seed-bank, providing the weeds do not pose a serious competition with the crop plants. Many of the plant species tested were not attacked, suggesting a different mechanism of resistance or lack of stimulants for *O. ramosa* seed germination. Of the species reported to be good “trap crops” for this parasite, *L. usitatissimum* showed very low infection, but *Brassica campestris*, *C. frutescens*, *C. annuum*, *Coriandrum sativum*, and *T. foenum-graecum* were heavily infected.

Orobanche ramosa is the most prevalent and devastating species to vegetables in irrigated and rain-fed agriculture, attacks a wide range of wild and cultivated plant species of different families (Qasem and Foy, 2007; Qasem, 2009) and is most destructive to tomato (*Lycopersicon esculentum* Mill), the main vegetable crop grown in the country. *Orobanche cernua* was the second-most commonly reported (parasitizing 20 species) and both parasitized the wild and cultivated plants of different plant families. *Orobanche palaestina*, *O. minor*, and *Orobanche schultzii* showed a more limited host range, with *O. schultzii* found parasitizing economically important crops in Jordan for the first time. The most frequently attacked families were: Compositae, Leguminosae, Solanaceae, and Umbelliferae by *O. ramosa*; Solanaceae and certain fruit trees of Moraceae, Oleaceae, Punicaceae, and Rosaceae by *O. cernua*; Compositae and Solanaceae by *O. aegyptiaca*; and Leguminosae by *O. crenata*. Woody shrubs and trees were attacked only by *O. cernua*, *O. palaestina*, and *O. schultzii*, while *O. minor* was only recorded on *V. faba* and *Daucus carota*. The intensity of infection (based on the number of emerged *Orobanche* shoots observed close to the host plants) greatly varied between the host species. Certain hosts were heavily infected, while others showed moderate or slight infection (Table 1.1.5). The heavily attacked species were: *Lycopersicon esculentum*, *Cucurbita pepo*, *Lactuca sativa*, *Lens culinaris*, *Malva sylvestris*, *Cucumis sativus*, and *Solanum melongena* by *O. ramosa*; *M. sylvestris*, *L. sativa*, *L. esculentum*, *S. melongena*, and *Solanum tuberosum* by *O. aegyptiaca*; *L. esculentum*, *Nicotiana tabacum*, and *S. melongena* by *O. cernua*, and *D. carota*, *Pisum sativum*, *L. esculentum*, and *V. faba* by *O. crenata*. The severity of infection varied within and between the parasite species depending on the dominating *Orobanche* species in the study site, the host species, and the prevailing environmental conditions, especially the temperature and soil type.

Orobanche ramosa is dominant in the central and northern parts of the Jordan Valley, while *O. crenata* is most common on *L. esculentum* in the southern part of the Jordan Valley and on *V. faba* in different locations. However, both *O. ramosa* and *O. cernua* are most common in the Jordan Valley, as well as in the highlands of the Mediterranean biogeographical region, while *O. cernua* is most common on cultivated vegetables under irrigation in marginal areas, the deserts, and dry regions. Parasitic species display host preferences in that the abundance of particular species was determined, for the most part, by the dominance of certain favorite hosts in vegetation at the sampled locations. Certain species of *Orobanche* (*O. cernua*, *O. palaestina*, and *O. schultzii*) attacked some fruit trees in addition to herbaceous species. However, our studies clearly showed that a wide host range of plants of different economic values are highly susceptible to different *Orobanche* species and that the infection on certain hosts is severe and destructive (*e.g. O. ramosa* on *L. esculentum* and *C. pepo*, *O. cernua* on *N. tabacum* and fruit trees, *O. crenata* on *D. carota* and *V. faba*). It also has been shown that many of the common weeds are hosts for certain parasitic species, perhaps guarding them from certain control measures by offering a sanctuary in which to produce and disperse their seeds.

In summary, broomrape is a root holo-parasitic flowering plant, lacks chlorophyll and parasitizes a wide range of wild and cultivated broadleaf crops worldwide. Seven species occur in agricultural lands in Jordan, attacking 86 plant species of different plant families, among which 50 species are cultivated crops. The parasite attaches itself to the root system of host species, sucking water, nutrients, and carbohydrates for the host plant directly through an absorptive organ called "haustorium". The parasite causes severe growth damage and, in most cases, collapsed host species under stress conditions and heavy infestation. Host species ranged from small herbaceous to high woody fruit and forest trees. The most common species is the branched broomrape "*Orobanche ramosa*" that attacks a wide range of species from Solanaceae, Cucrbitaceae, Cruciferea and Brassicaceae families in addition to many other economic crops of other plant families. *Orobanche* spp. are prolific seed producer widely spread in temperate regions under both irrigated and rainfed conditions, cause severe yield loss and may totally destroy yield and resulted in crop failure in many cases under both plastichouses and in the open field in Jordan. Conventional methods of *Orobanche* control are often practiced by local farmers in the absence of selective and effective herbicides. Manual pulling, hoeing, plastic mulching, planting date planning (mainly for tomato), chicken manure application, soil fumigation (in protected agriculture only), crop rotation, and soil solarization (in the Jordan Valley), are occasionally implemented in certain locations. Information on *Orobanche* spp. spread in Jordan, illustrations showing these parasites and some of their hosts in Jordan are shown (plates 1-135), and their methods of control under local conditions are presented.

Orobanchaceae spp. and host numbers for each in Jordan

Orobanche includes 7 species, attacks 86 host species (50 cultivated) of 24 families.

O. ramosa	73 host species
O. cernua	20 species
O. aegyptiaca	12 species
O. crenata	7 species
O. palaestina	6 species
O. minor	2 species
O. schultzii	2 species

Table 1.1.5. *Orobanche* species, hosts and their families, status and intensity of infection and prevalence in Jordan.

Parasite species/host	Family	Status	Intensity of infection	Prevalence
***a. Orobanche ramosa* L.**				
Allium cepa L.	Liliaceae	C	Light	Rare
Ammi majus L.	Umbelliferae	W	Moderate	Sporadic
Anethum graveolens L.	Umbelliferae	C	Light	GH
Anthemis cotula L.	Compositae	W	Moderate	Sporadic
Apium graveolense L.	Umbelliferae	C	Moderate	GH
Brassica oleracea L. *var. Botrytis*	Cruciferae	C	Light	Limited
Brassica oleracea L. *var. Capitata*	Cruciferae	C	Light	Limited
Brassica oleracea L.var. *caulorapa*	Cruciferae	C	Moderate	GH
Calendula officinalis L.	Compositae	C	High	Rare
Capsicum annuum L.	Solanaceae	C	Moderate	Sporadic
Capsicum frutescense L.	Solanaceae	C	Moderate	Limited
Cardaria draba (L.) Desv.	Cruciferae	W	Light	Rare
Carthamus tinctorius L.	Compositae	C	Moderate	Rare
Cichorium intybus L.	Compositae	W	Moderate	Sporadic
Citrullus vulgaris Schrad.	Cucurbitaceae	C	Light	Rare
Coleus blumei Benth.	Labiatae	C	Light	Rare
Coriandrum sativum L.	Umbelliferae	C	Light	Sporadic
Crepis aspera L.	Compositae	W	Light	Rare
Cucumis melo L.	Cucurbitaceae	C	Light	Limited
Cucumis melo L.*var. flexosus*	Cucurbitaceae	C	Light	Rare
Cucumis sativus L.	Cucurbitaceae	C	Light	Common
Cucurbita pepo L.	Cucurbitaceae	C	High	Very common
Dahlia pinnata Cav.	Compositae	C	Light	Rare
Datura metel L.	Solanaceae	W	Moderate	GH
Datura stramonium L.	Solanaceae	W	Light	GH
Daucus carota L.	Umbelliferae	C	Light	Limited
Dianthus caryophyllus L.	Caryophyllaceae	C	Light	Rare
Eruca sativa Miller	Cruciferae	W	Moderate	Sporadic
Eryngium creticum L.	Umbelliferae	W	Light	Sporadic
Falcaria vulgaris (L.) Karst.	Umbelliferae	W	High	Limited
Foeniculum vulgare Mill.	Umbelliferae	C	High	Common
Galium tricorne Stokes	Rubiaceae	W	Light	Rare
Gazania splendens Kiss. Mix.	Compositae	C	Moderate	Rare
Lactuca sativa L.	Compositae	C	High	Common
Lens culinaris Medik.	Leguminosae	C	High	Limited
Lepidium sativum L.	Cruciferae	C	Light	Limited
Linum usitatissimum L.	Linaceae	C	Light	GH
Lycopersicon esculentum Mill.	Solanaceae	C	High	Very common
Malva sylvestris L.	Malvaceae	W	High	Sporadic
Matricaria chamomilla L.	Compositae	W	Light	Limited
Matthiola annua (L.) Sweet	Cruciferae	C	Moderate	GH

Medicago sativa L.	Leguminosae	C	Light	Rare
Melilotus indicus (L.) All.	Leguminosae	W	Light	Sporadic
Mesembryanthemum sp.	Aizoaceae	C	Light	GH
Nicotiana tabacum L.	Solanaceae	C	Light	Limited
Ocimum basilicum L.	Labiatae	C	Light	Sporadic
Oxalis corniculata L.	Oxalidaceae	W	Moderate	GH
Pelargonium grandiflorum (Andrews) Willd.	Geraniaceae	C	Light	Rare
Pelargonium zonale (L.) Aiton	Geraniaceae	C	Moderate	GH
Petunia x hybrid Vilm	Solanaceae	C	Moderate	GH
Picnomon acarna (L.) Cass	Compositae	W	Light	Rare
Polygonum aviculare L.	Polyginaceae	W	Moderate	Limited
Punica granatum L.	Punicaceae	C	Light	GH
Rhagadiolus stellatus (L.) Willd.	Compositae	W	Light	Sporadic
Rosa damascena Mill.	Rosaceae	C	Light	GH *Rumex*
acetosella L.	Polygonaceae	W	Moderate	Limited
Salvia splendens Ker- Gawl.	Labiatae	C	Moderate	GH
Silybum marianum (L.) Gaertn.	Compositae	W	Light	Sporadic
Solanum melongena L.	Solanaceae	C	High	Common
Solanum nigrum L.	Solanaceae	W	Moderate	Limited
Solanum tuberosum L.	Solanaceae	C	Moderate	Common
Thunbergia alata Bojer ex Sims	Acanthaceae	C	Moderate	GH
Thymus syriacus Boiss.	Labiatae	C	Moderate	GH
Tragopogon coelesyriacus Boiss.	Compositae	W	Light	Rare
Trifolium L.	Leguminosae	C	Moderate	Common
Tropaeolum majus L.	Tropaeolaceae	C	Moderate	GH
Verbena officinalis L.	Verbanaceae	W	Moderate	Sporadic
Vicia faba L.	Leguminosae	C	High	Very common
Vicia narbonensis L.	Leguminosae	W	Light	Sporadic
Vicia palestina Boiss.	Leguminosae	W	Light	Sporadic
Vicia peregrina L.	Leguminosae	W	Light	Rare
Withania somnefera Dunal	Solanaceae	W	Light	Rare
Xanthium strumarium L.	Compositae	W	Light	Sporadic
***b. Orobanche aegyptiaca* Pers.**				
Brassica oleracea L. *var. capitata*	Cruciferae	C	Moderate	Limited
Pelargonium grandiflorum (Andrews) Willd.	Geraniaceae	C	Light	GH
Gerapogon hybridus (L.) Sch.	Compositae	W	Light	Rare
Lactuca sativa L.	Compositae	C	High	Common
Lycopersicon esculentum Mill	Solanaceae	C	High	Very common
Malva sylvestris L.	Malvaceae	W	High	Sporadic
Prunus persica (L.) Batsch	Rosaceae	C	Light	Sporadic
Solanum melongena L.	Solanaceae	C	High	Common
Solanum tuberosum L.	Solanaceae	C	High	Limited
Sonchus oleraceous L.	Compositae	W	Light	Rare
Vicia faba L.	Leguminosae	C	Light	Limited
Xanthium spinosum L.	Compositae	W	Light	Rare
***c. Orobanche cernua* Loefl.**				

Amygdalus communis L.	Rosaceae	C	High	Limited
Cucumis melo L.	Cucurbitaceae	C	Moderate	Limited
Daucus carota L.	Umbelliferae	C	Moderate	Common
Ficus carica L.	Moraceae	C	Moderate	Limited
Helianthus annuus L.	Compositae	C	Light	Rare
Lactuca orientalis Boiss.	Compositae	W	Light	Sporadic
Lycopersicon esculentum Mill.	Solanaceae	C	High	Very common
Nicotiana tabacum L.	Solanaceae	C	High	Very common
Olea europaea L.	Oleaceae	C	Light	Limited
Petunia hybrida Vilm	Solanaceae	C	Light	GH
Prunus armeniaca L.	Rosaceae	C	Moderate	Rare
Prunus persica (L.) Batsch	Rosaceae	C	Light	Sporadic
Punica granatum L.	Punicaceae	C	Light	GH
Quercus coccefera L.	Fagaceae	C	Light	Limited
Silybum marianum (L.) Gaertn	Compositae	W	Light	Sporadic
Solanum melongena L.	Solanaceae	C	High	Limited
Solanum tuberosum L.	Solanaceae	C	Light	Common
Salvia splendens Ker Gawl.	Labiatate	C	High	GH
Vicia faba L.	Leguminosae	C	Light	Common
Xanthium spinosum L.	Compositae	W	Light	Rare
***d. Orobanche crenata* Forskl.**				
Ammi majus L.	Umbelliferae	W	Light	Limited
Daucus carota L.	Umbelliferae	C	High	Very common
Lycopersicon esculentum Mill.	Solanaceae	C	high	Very common
Melilotus indicus (L.) All.	Leguminosae	W	Moderate	Sporadic
Pisum sativum L.	Leguminosae	C	High	Limited
Rhagadiolus stellatus (L.) Willd	Compositae	W	Light	Sporadic
Vicia faba L.	Leguminosae	C	High	Very common
***e. Orobanche palaestina* Reut.**				
Amygdalus communis L.	Rosaceae	C	Moderate	Sporadic
Cirsium syriacum (L.) Gaertn	Compositae	W	Light	Rare
Daucus carota L.	Umbelliferae	C	Light	Rare
Eryngium creticum L.	Umbelliferae	W	Light	Rare
Olea europaea L.	Oleaceae	C	High	Rare
Quercus coccefera L.	Fagaceae	W	Moderate	Rare
Rhagadiolus stellatus (L.) Willd.	Compositae	W	Light	Rare
***f. Orobanche minor* Sm.**				
Daucus carota L.	Umbelliferae	C	Moderate	Very common
Vicia faba L.	Leguminosae	C	Moderate	Common
***g. Orobanche schultzii* Mutel**				
Amygdalus communis L.	Rosaceae	C	Moderate	Limited
Olea europaea L.	Oleaceae	C	High	Limited

C = Cultivated; W = Wild; GH = Glasshouse grown specie; NC = Attachment was not confirmed

Rare: only on few plants in 1-2 sites of a biogeographical region; Sporadic: few plants infected in one or more biogeographical regions; Limited: on many plants localized in certain locations of 1 or 2 biogeographical regions. Common: on certain plant species in > one biogeographica

Very common: on many plant species in different locations of different biogeographical regions

Source: modified after Qasem and Foy, 2007; Qasem, 2009; Qasem, 2011

Table 1.1.6. Distribution of hosts between *Orobanche* spp. found in Jordan.

Family	Host plant Species	*O. aegyptiaca*	*O. cernua*	*O. crenata*	*O. minor*	*O. rpalaestina*	*O. ramosa*	*O. schultzii*
Acanthaceae	*Thunbergia alata*						X	
Aizoaceae	*Mesembryanthemum* sp.						X	
Caryophyllaceae	*Dianthus caryophyllus*		X					
Compositae	*Anthemis cotula*						X	
	Calendula officinalis						X	
	Carthamus tinctorius.						X	
	Cichorium intybus						X	
	Cirsium syriacum					X	X	
	Crepis aspera						X	
	Dahlia pinnata						X	
	Gazania splendens						X	
	Gerapogon hybridus	X						
	Helianthus annuus		X					
	Lactuca orientalis		X					
	Lactuca sativa	X					X	
	Matricaria chamomilla						X	
	Picnomon acrana						X	
	Rhagadiolus stellatus			X		X	X	
	Silybum marianum.		X				X	
	Sonchus oleraceus.	X						
	Tragopogon coelesyriacus						X	
	Xanthium spinosum	X	X					
	Xanthium strumarium						X	
Cruciferae	*Brassica oleracea var. botrytis*						X	
	Brassica oleracea var. capitata	X					X	
	Brassica oleracea L. var. *caulorapa*						X	
	Cardaria draba						X	
	Eruca sativa						X	
	Lepidium sativum						X	
	Matthiola annua						X	
Cucurbitaceae	*Citrullus vulgaris*						X	
	Cucumis melo		X				X	
	Cucumis melo var. flexosus						X	
	Cucumis sativus						X	
	Cucurbita pepo						X	
Fagaceae	*Quercus coccifera*						X	
Geraniaceae	*Pelargonium grandiflorum*	X				X	X	
	Pelargonium zonale						X	
Labiatae	*Coleus blumei*						X	

	Ocimum basilicum						X	
	Salvia splendens		X				X	
	Thymus syriacus						X	
Leguminosae	*Lens culinaris*						X	
	Medicago sativa						X	
	Melilotus indicus			X			X	
	Pisum sativum.			X				
	Trifolium hybridum						X	
	Vicia faba	X	X	X	X		X	
	Vicia narbonensis						X	
	Vicia palestina.						X	
	Vicia peregrine						X	
Liliaceae	*Allium cepa*						X	
Linaceae	*Linum usitatissimum*						X	
Malvaceae	*Malva sylvestris*	X					X	
Moraceae	*Ficus carica*		X					
Oleaceae	*Olea europaea.*		X			X		X
Oxalidaceae	*Oxalis corniculata.*						X	
Polyginaceae	*Polygonum aviculare.*						X	
	Rumex acetosella						X	
Punicaceae	*Punica granatum*		X				X	
Rosaceae	*Amygdalus communis*		X			X		X
	Prunus armeniaca		X					
	Prunus persica	X	X					
	Rosa hybrida						X	
Rubiaceae	*Galium tricorne*						X	
Solanaceae	*Capsicum annuum*						X	
	Capsicum frutescence						X	
	Datura metel						X	
	Datura stramonium						X	
	Lycopersicon esculentum	X	X	X			X	
	Nicotiana tabacum		X				X	
	Petunia hybrida		X				X	
	Solanum melongena	X	X				X	
	Solanum nigrum						X	
	Solanum tuberosum	X	X				X	
	Withania somnifera						X	
Tropaeolaceae	*Tropaeolum majus*						X	
Umbelliferae	*Ammi majus*			X			X	
	Anethum graveolens						X	
	Apium graveolens						X	
	Coriandrum sativum						X	
	Daucus carota		X	X	X	X	X	
	Eryngium creticum					X	X	
	Falcaria vulgaris						X	
	Foeniculum vulgare						X	
Verbanaceae	*Verbena officinalis*						X	

X = infected,

Source: Qasem and Foy, 2007; Qasem, 2009; Qasem, 2011

SOME ILLUSTRATIONS ON *OROBANCHE* SPP. AND THEIR HOSTS IN JORDAN

Orobanche aegyptiaca

Plate 1. *Orobanche aegyptiaca* on common mallow (*Malva sylvestris*)

Plate 2. *Orobanche aegyptiaca* showed attached on *Malva sylvestris* roots

Plate 3. *Orobanche aegyptiaca* on *Carthamus nitidus*

Plate 4. *Orobanche aegyptiaca* on *Mercurialis annua*

Plate 5. *Orobanche aegyptiaca* on lentil (*Lens esculentus*)

Plate 6. *Orobanche aegyptiaca* on the tuber of potato (*Solanum tuberosum*) as appeared in the field

Plate 7. *Orobanche aegyptiaca* on potato (*Solanum tuberosum*) tuber

Plate 8. *Orobanche aegyptiaca* attached to the potato roots and tubers

Plate 9. *Orobanche aegyptiaca* on faba bean (*Vicia faba*)

Orobanche crenata

Plate 10. *Orobanche crenata* and *O. ramosa* attached to roots of *Ammi majus*

Plate 11. *Orobanche crenata* attached melilot (*Melilotus indicus*) roots

Plate 12. *Orobanche crenata* on *Astragalus* sp.

Plate 13. *Orobanche crenata* on *Pelargonium zonale*

Plate 14. *Orobanche crenata* on *Petumia medium*

Plate 15. *Orobanche crenata* infestation on broadbean (*Vicia faba*)

Plate 16. *Orobanche crenata* killed broabean (*Vicia faba*) plants

Plate 17. *Orobanche crenata* eliminating broadbean plants

Plate 18. *Orobanche crenata* infestation on carrots (*Daucus carota*)

Plate 19. *Orobanche crenata* attachments on the roots of *Daucus carota*

Plate 20. *Orobanche crenata* on fennel (*Foeniculum vulgare*)

Plate 21. *Orobanche crenata* on peas (*Pisum sativum*) in pots under glasshouse conditions

Plate 22. *Orobanche crenata* parasitizing peas (*Pisum sativum*) in the field

Plate 23. *Orobanche crenata* infestation on tomato (*Lycopersicon esculentum*)

Plate 24. *Orobanche crenata* infestation in tomato field

Orobanche cernua

Plate 25. *Orobanche cernua* on *Notobasis syriaca*

Plate 26. *Orobanche cernua* on *Pelargonium roseum*

Plate 27. *Orobanche cernua* on musk melon (*Cucumis melo*)

Plate 28. *Orobanche cernua* and *Orobanche ramosa* on potato

Plate 29. *Orobanche cernua* parasitizing sunflower (*Helianthus annuus*) in the field

Plate 30. *Orobanche cernua* on tobacco (*Nicotiana tabacum*)

Plate 31. *Orobanche cernua* on tomato

Plate 32. *Orobanche cernua* infesting tomato field

Plate 33. *Orobanche cernua* on eggplant (*Solanum melongena*)

Plate 34. *Orobanche cernua* on almond (*Amygdalus communis*)

Plate 35. *Orobanche cernua* on wild almond

Plate 36. *Orobanche cernua* on olive (*Olea europaea*)

Plate 37. *Orobanche cernua* on *Prunus persica*

Plate 38. *Orobanche cernua* attacked fig (*Ficus carica*)

Plate 39. *Orobanche cernua* close to grape (*Vitis vinifera*)

Plate 40. *Orobanche cernua* on peach (*Prunus domestica*)

Plate 41. *Orobanche cernua* on *Eriobotrya japonica* (Thunb.) Lindl

Plate 42. *Orobanche cernua* on pomegranate (*Punica granatum*)

Plate 43. *Orobanche cernua* on sumac (*Rhus coriaria*)

Orobanche minor

Plate 44. *Orobanche minor* on carrot (*Daucus carota*)

Plate 45. *Orobanche minor* on broadbean (*Vicia faba*)

Orobanche ramosa

Plate 46. *Orobanche ramosa* on *Cichorium intybus*

Plate 47. *Orobanche ramosa* on *Anthemis cotula*

Plate 48. *Orobanche ramosa* on *Cardaria draba*

Plate 49. *Orobanche ramosa* on *Eruca sativa*

Plate 50. *Orobanche ramosa* on *Peganum harmala*

Plate 51. *Orobanche ramosa* on *Carduus pycnocephalus*

Plate 52. *Orobanche ramosa* on *Coronilla scorpioides*

Plate 53. *Orobanche ramosa* on *Verbena* sp.

Plate 54. *Orobanche ramosa* on *Gerapogon hybridus*

Plate 55. *Orobanche ramosa* on medic (*Medicago sativa*)

Plate 56. *Orobanche ramosa* on *Melilotus alba*

Plate 57. *Orobanche ramosa* on *Solanum elaeagnifolium*

Plate 58. *Orobanche ramosa* on *Falcaria vulgaris*

Plate 59. *Orobanche ramosa* on *Falcaria vulgaris* at early growth stage

Plate 60. *Orobanche ramosa* on *Rumex dentatus*

Plate 61. *Orobanche ramosa* on *Solanum nigrum*

Plate 62. *Orobanche ramosa* on *Hyoscyamus* sp

Plate 63. *Orobanche ramosa* infection on *Oxalis corniculata*

Plate 64. *Orobanche ramosa* attached on *Oxalis corniculata*

Plate 65. *Orobanche ramosa* on *Polygonum aviculare*

Plate 66. *Orobanche ramosa* on *Vicia palaestina*

Plate 67. *Orobanche ramosa* on *Vicia syriaca*

Plate 68. *Orobanche ramosa* on *Foeniculum vulgare*

Plate 69. *Orobanche ramosa* attached to fennel (*Foeniculum vulgare*) root

Plate 70. *Orobanche ramosa* on *Carum carvi*

Plate 71. *Orobanche ramosa* on *Melissa officinalis*

Plate 72. *Orobanche ramosa* on squash (*Cucurbita pepo*) under plastic houses

Plate 73. *Orobanche ramosa* on squash (*Cucurbita pepo*) in the field

Plate 74. *Orobanche ramosa* on cucumber (*Cucumis sativus* L.)

Plate 75. *Orobanche ramosa* at floweing stage on tomato

Plate 76. *Orobanche ramosa* at seeding stage on tomato

Plate 77. *Orobanche ramosa* on tomato showing different flower color

Plate 78. *Orobanche ramosa* on *Solanum tuberosum*

Plate 79. *Orobanche ramosa* on *Anethum graveolens*

Plate 80. *Orobanche ramosa* parasitizing ajwan (*Trachyspermum ammi*)

Plate 81. *Orobanche ramosa* on safflower (*Carthamus tinctorius*)

Plate 82. *Orobanche ramosa* showing whitish flowers on safflower *Carthamus tinctorius*

Plate 83. *Orobanche ramosa* on *Nigella damascene*

Plate 84. *Orobanche ramosa* extremely low infestation on flax (*Linum usitatissimum* L.).

Plate 85. *Orobanche ramosa* on *Thymus syriacus*

Plate 86. *Orobanche ramosa* attached on roots of *Anethum graveolens*

Plate 87. *Orobanche ramosa* on *Helianthus annuus* drying the host and failed flowering

Plate 88. *Orobanche ramosa* on ornamental *Helianthus annuus*

Plate 89. *Orobanche ramosa* on basil (*Ocimum basilicum*)

Plate 90. *Orobanche ramosa* on different cultivar of basil (*Ocimum basilicum*)

Plate 91. *Orobanche ramosa* on soybean (*Glycine max*)

Plate 92. *Orobanche ramosa* on watermelon (*Citrullus vulgaris*)

Plate 93. *Orobanche ramosa* on purslan (*Portulaca oleracea*)

Plate 94. *Orobanche ramosa* on anis (*Pimpinella anisum*)

Plate 95. *Orobanche ramosa* on anis (*Pimpinella anisum*)

Plate 96. *Orobanche ramosa* on parsley (*Petroselinum sativum*)

Plate 97. *Orobanche ramosa* on cumin (*Cuminum cyminum*)

Plate 98. *Orobanche ramosa* on plastic-mulched eggplant

Plate 99. *Orobanche ramosa* cause wilting to eggplant

Plate 100. *Orobanche ramosa* on spinach (*Spinacea oleracea*)

Plate 101. *Orobanche ramosa* on hot pepper (*Capsicum annuum*)

Plate 102. *Orobanche ramosa* attached to hot pepper roots

Plate 103. *Orobanche ramosa* attached to cabbage (*Brassica oleracea* L. *var. capitata*) roots

Plate 104. *Orobanche ramosa* very weakly connected on onion (*Allium cepa*) roots

Plate 105. *Orobanche ramosa* on *Tropaeolum majus nanum*

Plate 106. *Orobanche ramosa* on *Coleus hybridus*

Plate 107. *Orobanche ramosa* on *Chrysanthemum carinatum*

Plate 108. *Orobanche ramosa* showing different flower color on *Chrysanthemum carinatum*

Plate 109. *Orobanche ramosa* on *Zennia elegans*

Plate 110. *Orobanche ramosa* on *Limonium sinuatum* (L.) Mill

Plate 111. *Orobanche ramosa* on *Brassica oleracea var. acephala*

Plate 112. *Orobanche ramosa* on *Dahlia variabilis* (Willd) Desf.

Plate 113. *Orobanche ramosa* on *Mesembryanthemum criniflorum*

Plate 114. *Orobanche ramosa* on *Salvia splendens*

Plate 115. *Orobanche ramosa* on *Convolvulus tricolor* (syn. *C. minor*)

Plate 116. *Orobanche ramosa* on *Petunia hybrida*

Plate 117. *Orobanche ramosa* on *Centaurea* sp.

Plate 118. *Orobanche ramosa* on *Vinca rosea* L.

Plate 119. *Orobanche ramosa* on initiated roots of pomegranate (*Punica granatum* L.) stem cut

Orobanche palaestina

Plate 120. *Orobanche palaestina* on weeds

Plate 121. *Orobanche palaestina* on almond (*Amygdalus communis*)

Plate 122. *Orobanche palaestina* on olive (*Olea europaea*)

Plate 123. *Orobanche palaestina* attached on oak (*Quercus coccefera*) roots

Plate 124. *Orobanche palaestina* infestation on oak (*Quercus coccefera*)

Plate 125. *Orobanche palaestina* on sumac (*Rhus coriaria*)

Plate 126. *Orobanche palaestina* on peach (*Prunus domestica*)

Plate 127. *Orobanche palaestina* on *Prunus persica*

Plate 128. *Orobanche palaestina* on apricot (*Prunus armeniaca*)

Plate 129. *Orobanche palaestina* on Canada red chokecherry (*Prunus virginiana*)

Plate 130. *Orobanche palaestina* on Canada red chokecherry (*Prunus virginiana*) growing close to the crown area

Plate 131. *Orobanche palaestina* under *Citrus* sp.

Orobanche schultzii

Plate 132. *Orobanche schultzii* on *Cirsium phyllocephalum*

Plate 133. *Orobanche schultzii* on olive at haustorial stage to complete emerged parasite

Plate 134. *Orobanche schultzii* at flowering stage on olive

Plate 135. *Orobanche schultzii* on almond

REFERENCES

Abbes, Z., Trabelsi, I., Kharrat, M., Amri, M. Intercropping with fenugreek (*Trigonella foenum-graecum*) enhanced seed yield and reduced *Orobanche foetida* infestation in faba bean (*Vicia faba*). *Biol. Agric. Hortic.*, 2019, 35, 4, 238-247, DOI: 10.1080/01448765.2019.1616614.

Abdalla, M. M. F. No tillage sweet corn production following faba bean and effect on *Orobanche* seeding depth. *Bulletin of Faculty of Agriculture, University of Cairo*, 1999, 50, 416-435.

Abdalla, N. E. K., Dabrowski, Z.T. Effect of soil solarization on Orobanche in the central Sudan. *The FAO/ARC IPM Project GCP/SUS/025/NET*, P.O. Box 196. Wad Medani, Sudan. 2000.

Abdel-Rahim, M. F., Satour, M. M., Mickail, K. Y., El-Eraki, S. A., Grinstein, A., Chen, Y., Katan, J. Effectiveness of soil solarization in furrow-irrigated Egyptian soils. *Plant Dis.*, 1988, *72*, 143-146.

Abdel-Kader, M. M., El-Mougy, N. S. Applicable control measure against *Orobanche ramosa* in tomato plants. *Australas. Plant Pathol.*, 2007, 36, 160–164.

Abdel-Kader, M. M., El-Mougy, N. S. Prospects of mycoherbicides for control of broomrapes (*Orobanche* spp.) in Egypt. *J. Plant Prot. Res.*, 2009, 49 (1), 63-75.

Abdel-Kader, M.M., Badiaa, R., Ismail, R., Diab, M.M. Preliminary Evaluation of Some Soilborne Fungi Parasitizing *Orobanche crenata* in Greenhouse. In: Maillet, J., (ed.), *6th EWRS Mediterranean Symposium,* Montpellier, France, 1998, pp. 127-132.

Abebe, G., Sahile, G., Al-Tawaha, A.R. M. Evaluation of potential trap crops on *Orobanche* soil seed bank and tomato yield in the central rift valley of Ethiopia. *World Journal of Agricultural Sciences*, 2005, 1 (2), 148-151.

Abedi, S., Darvishzadeh, R., Bernousi, I., Mandoulakani, B.A., Maleki, H.H., Shah, D. Genetic variability of *Orobanche aegyptiaca* infesting tobacco in Iran by Bayesian analysis. *Biologia*, 2014, 69/12, 1652—1659, DOI: 10.2478/s11756-014-0473-6.

Abou-Salama, A. M. Utilization of crop rotation for the control of *Orobanche crenata* Forsk. *Assiut Journal of Agricultural Sciences*, 1995, 26, 245-252.

Abu-Irmaileh, B.E. Soil solarization controls broomrapes (*Orobanche* sp.) in host vegetable crops in the Jordan valley. *Weed Technol.*, 1991, 5, 575- 581.

Abu-Shakra, S., Maih, A.A., Saghir, A.R. Germination of seeds of branched broomrape (*Orobanche ramosa* L.). *Hortic. Res.*, 1970, 10, 119-124.

Acharya, B. D. Assesment of different non-host crops as trap crop for reducing *Orobanche aegyptiaca* Pers. Seed Bank. *Ecoprint.*, 2012, 19, 31-38.

Acharya, B. D., Khattri, G. B., Chettri, M. K., Srivastava, S. C. Effect of *Brassica campestris* var. *toria* as a catch crop on *Orobanche aegyptiaca* seed bank. *Crop Prot.,* 2002, 21, 533-537.

Agabawi, K.M., Younis, A.E. Effect of nitrogen application on growth and nitrogen content of *Striga hermonthica* (Del.) Benth. and *Sorghum vulgare* Lur. grown for forage. *Plant Soil.* , 1985, 23, 295-304.

Akhtouch, B., del Moral, L., Leon, A., Velasco, L., Fernández-Martínez, J.M., Pérez-Vich, B. Genetic study of recessive broomrape resistance in sunflower. *Euphytica*, 2016; 209, 419–428.

Aksoy, E., Arslan, Z.F., Tetik, Ö., Eymirli, S. Using the possibilities of some trap, catch and Brassicaceae crops for controlling crenate broomrape a problem in lentil fields. *Int. J. Plant Prod.*, 2016, 10 (1), 53-62.

Alakonya, A., Kumar, R., Koenig, D., Kimura, S., Townsley, B., Runo, S., Garces, H.M., Kang, J., Yanez, A., David-Schwartz, R., Machuka, J., Sinha, N. Interspecific RNA interference of shoot meristemless-like disrupts *Cuscuta pentagona* plant parasitism. *Plant Cell,* 2012, 24, 3153–3166.

Al-Khateeb, W. M. *Orobanche and Tomato Interaction as Influenced by Salinity and Water Deficit under in Vivo and in Vitro Conditions*. M.Sc. Thesis, Jordan University of Science and Technology, Irbid, Jordan. 2002.

Al-Khesraji, T.O., Abdel Wahib, A.U., Annou, M.R. Insect pests attacking parasitic flowering plants in the southern desert of Iraq. *Iraqi J. Agr. Sci. "Zanco"*, 1987, 5, 197–216.

Al-Menoufi, A.O. *Cultural Practices To Control Orobanche crenata in Faba Bean Field in Egypt. Final Report,* Project Financed by Alexandria University, Egypt (1991-1992).1992.

Al-Menoufi, A.O., Adam, M.A. Susceptibility/resistance of Some Wild *Lycopersicon* Accessions to *Orobanche ramosa*. In: Wegmann, K., Musselman, L.J., Joel, D.M., (eds.). *Current Problems of Orobanche Researches, Proceedings of the 4th International Orobanche Workshop*, Sep. 23-26, 1998, Institute for wheat and sunflower "Dobroudja", Albena, Bulgaria, 1998, pp. 259-266.

Al-Menoufi, O.A. Crop rotation as a control measure of *Orobanche crenata* in *Vicia faba* fields. In: Wegmann, K., Musselman, L.J., (eds.), *Progress in Orobanche Research. Proceedings, International Workshop on* Orobanche *Research*, Obermarchtal, 1989. Eberhard-Karls-Universitat, Tubingen. 1991, pp. 241-247.

Al-Menoufi, O.A., Adam, M.A. Biological and chemical inhibition of *Orobanche* seed germination. In: Moreno, M.T., Cubero, J.I., Derner, D., Joel, D., Musselman, L.J., Parker, C., (eds.). *Advances in Parasitic Plant Research, Sixth International Parasitic Weed Symposium*, April 16-18, 1996, Cordoba, Spain. 1996, pp. 418-423.

Alonso L.C. Resistance to *Orobanche* and resistance breeding: A review. In: Wegmann, K., Musselman, L.J., Joel, D.M., (eds.), *Current Problems of Orobanche Researches. Proceedings of the Fourth International Workshop on Orobanche,* Albena, 23-26, September, Bulgaria.1998, pp. 233-257.

Alonso, L.C., Rodriguez-Ojeda M.I., Fernandez-Escobar, J., Lopez-Ruiz-Calero, G. Chemical control of broomrape (*Orobanche cernua* Loefl.) in sunflower (*Helianthus annuus* L.) resistant to imazethapyr herbicide. *Helia*, 1998, 21, 45-53.

Aly, R. Conventional and biotechnological approaches for control of parasitic weeds. *In Vitro Cell. Dev. Biol. Plant*., 2007, 43,304–317. DOI 10.1007/s11627-007-9054-5

Aly, R., Granot, D., Mahler-Slasky, Y, Halpern, N., Galun, E. Saccharomyces cerevisiae cells, harboring the gene encoding Sarcotoxin IA secrete a peptide that is toxic to plant pathogenic bacteria. *Protein Expr. Purif.*, 1999, 16,120–124.

Aly, R., Plakhin, D., Achdari, G. Expression of sarcotoxin IA gene via a root-specific tob promoter enhanced host resistance against parasitic weeds in tomato plants. *Plant Cell Rep.*, 2006, 25, 297–303.

Aly, R.; Goldwasser, Y., Eizenberg, H., Hershenhorn, J., Golan, S., Kleifeld, Y. Broomrape (*Orobanche cumana*) control in sunflower (*Helianthus annuus*) with imazapic. *Weed Technol.*, 2001, 15, 306-309.

Americanos, P.G. *Control of* Orobanche *in broadbeans. Technical Bulletin*, Agricultural Research Institute, Nicosia, Cyprus. No. 1983, 50, 4pp.

Americanos, P.G. *Control of* Orobanche *in Celery. Technical Bulletin* 137, Agricultural Research Institute and Ministry of Agriculture and Natural Resources. Nicosia, Cyprus, 1991.

Americanos, P.G., Vouzounis, N.A.. *Control of* Orobanche *in cabbage.* Nicosia, Cyprus; Agricultural Research Institute, *Cyprus Technical Bulletin-* Cyprus Agricultural Research Institute, 1995, 170, 7PP.

Amsellem, Z., Barghouthi, S., Cohen, B., Goldwasser, Y., Gressel, J., Hornok, L., Kerenyi, Z., Kleifeld, Y., Klein, O., Kroschel, J., Sauerborn, J., Muller-Stover, D., Thomas, H., Vurro, M.,

Zonno, M. C. Recent advances in the biocontrol of *Orobanche* (broomrape) species. *BioControl*, 2001, 46, 211-228.

Amsellem, Z., Zidack, N.K., Quimby, Jr, P.C., Gressel, J. Long term dry preservation of active mycelia of two mycoherbicidal organisms. *Crop Prot*., 1999 (10), 18, 643–649.

Antonova T.S., Borg S.J.Ter. The role of peroxidase in the resistance of sunflower against *Orobanche cumana* in Russia. *Weed Res*., 1996, 36 (2), 113-121.

Aregawi, T. B. *The Synergistic Effect of Host Resistance, Rhizobial inoculant and* Trichoderma *on* Orobanche *Infestation of Faba Beans* (Vicia faba L.) *in Southen Tigray, Ethiopia.* M.Sc. Thesis. Hawassa University, Ethiopia. 2017.

Arjona-Berral, A., Garcia-Torres, L. Broomrape control in lentils with glyphosate. *LENS*. 1983, 10, 20-22.

Auger, B.; Pouvreau, J.B., Pouponneau, K., Yoneyama, K., Montiel, G., Le Bizec, B., Yoneyama, K., Delavault, P., Delourme, R., Simier, P. Germination stimulants of *Phelipanche ramosa* in the rhizosphere of *Brassica napus* are derived from the glucosinolate pathway. *Mol. Plant Microbelnteract.* 2012, 25, 993–1004.doi:10.1094/MPMI-01-12- 0006-R.

Auld, B.A., Morin, L. Constraints in the development of bioherbicides. *Weed Technol.,* 1995, 9, 638-652.

Avdeev, Y. I., Shcherbinin, B.M. Investigating tomato resistance to broomrape and the inheritance of this trait. *Daklady Vsesoyuznoi Akademii Sel'skokhozyaistevennykh Nauk imeni V.I.Lenina*., 1978, 1, 14-16.

Avdeev, Y. I., Shcherbinin, B.M. Tomato resistance to the broomrape *Orobanche aegyptiaca* Pers. *Tsiologiya I Genetika*, 1977, 11, 491-496.

Avdeev, Y.I. Identification of a dominant gene for resistance to broomrape in tomato and its uses in breeding. *Plant Breeding Abstract*, 1986, 50, 3944.

Avdeev, Y.I. The variety 'Astrakhan' is resistant to broomrape. *Kartofel' I Ovoshchi*, 1998, No.2, 40.

Avdeev, Y.I., Scherbinin, B.M. Resistance of tomato to sunflower broomrape *Orobanche cumana*. *Plant Breeding Abstract*, 1983, 56, 2308

Aviv, D., Amsellem, Z., Gressel, J. Transformation of carrots with mutant acetolactate synthase for *Orobanche* (broomrape) control. *Pest Manag. Sci.*, 2002, 58, 1187–1193.

Aybeke, M. *Fusarium* infection causes genotoxic disorders and antioxidant-based damages in *Orobanche* spp. *Microbiol. Res.*, 2017a, 201, 46–51.

Aybeke, M. *Fusarium* infection causes phenolic accumulations and hormonal disorders in *Orobanche* spp. *Indian J Microbiol.,* 2017b, 57(4), 416–421. DOI 10.1007/s12088-017-0669-x

Aybeke, M. *Orobanche cernua* control experiments with different composting methods. *Acta Biologica Turcica.*, 2016, 29 (4), 128-136.

Aydin, A., Mutlu, H. Broomrape development on sunflower planted at different dates. *Helia*, 1996, 19, 105-110.

Aziz, Kh.M. The use of microelements and herbicides. *Zashchita Rastenii (Moskva)*. 1989, 10, 30.

Babaei, S., Alizadeh, H., Jahansouz, M. R., Mashhadi, H. R., Moeini, M.M. Management of *Phelipanche aegyptiaca* Pomel. using trap crops in rotation with tomato (*Solanum lycopersicom* L.). *Aust. J. Crop Sci.* (*AJCS*), 2010, 4(6), 437-442.

Babalola, O. O. Exogenous cellulase contributes to mycoherbicidal activity of *Fusarium arthrosporioides* on *Orobanche aegyptiaca. Int. J. Agron.*, 2010, Article ID 963259, 4 pages, 2010. Available at: https://doi.org/10.1155/2010/963259

Ballard, B.J. *The Effects of Various Root Exudates on Broomrape Germination.* 1979, MSc. Thesis, VPI, Blacksburg, Virginia, USA

Baloch, G.M., Din, I.M., Ghani, M.A. Parasitic weed plants and their natural enemies with special reference to Pakistan – a review. *Pakistan Journal of Forestry* (PJF), 1980, 30, 2–13.

Bandaranayake, P.C.G., Yoder, J.I. Trans-specific gene silencing of acetyl-CoA carboxylase in a root-parasitic plant. *Mol. Plant Microbe Interact.*, 2013, 26, 575–584.

Bardaro, N., Marcotrigiano, A.R., Bracuto, V., Mazzeo, R., Ricciardi, F., Lotti, C., Pavan, S., Ricciardi, L. Genetic analysis of resistance to *Orobanche crenata* (Forsk.) in a pea (*Pisum sativum* L.) low-strigolactone line. *J. Plant Pathol.*, 2016, 98, 671–675.

Bari, V.K., Nassar, J.A., Kheredin, S.M., Gal-On, A., Ron, M., Britt, A., Steele, D., Yoder, J., Aly, R. CRISPR/Cas9-mediated mutagenesis of carotenoid cleavage dioxygenase 8 in tomato provides resistance against the parasitic weed *Phelipanche aegyptiaca. Sci. Rep.*, 2019, 9, e11438.

Barney, L., Lipscomb, B. L., Diggs, Jr. G. M. *Hydrocotyle sibthorpioides* (Apiaceae) new for Texas and notes on introduced species. *SIDA, Contributions to Botany*, 2005, 21 (4), 2449-2456. The Botanical Research Institute of Texas, Inc. Available at: http://www-aes.tamu.edu/mary/brmrape/brmrape.htm

Bar-Nun, N., Mayer, A.M. Composition of and changes in storage compounds in *Orobanche aegyptiaca* seeds during preconditioning. *J. Plant Sci.*, 2002, 50(4), 277-279.

Batchvarova, R., Hristov, M., Slavov, S., Valkov, V. Development of sunflower lines resistant to *Orobanche cumana* Wallr through mutagenesis and interspecific hybridization. In: Fer, A., Thalouarn, P., Joel, D.M., Musselman, L.J., Parker, C., Verkleij, J.A.C., (eds.), *Proceedings of the 7th International Parasitic Weed Symposium*, Nantes, France. 2001, pp. 233-234.

Batchvarova, R.; Slavov, S., Valkov, V., Atanassova, S., Atanassova, A. Control of *Orobanche* spp. by herbicides resistant crops: an example with transgenic tobacco. In: Maillet, J. (ed.), *Comptes-rendus 6eme Symposium Mediterranean EWRS,* Montpellier, France, 13-15 mai, 1998, pp. 153-154.

Bedi, J. S. Further studies on control of sunflower broomrape with *Fusarium oxysporum* f. sp. *orthoceras*- a potential mycoherbicides. In: Pieterse A.H., Verkleij J.A.C., ter Borg., S.J., (eds.), *Biology and Management of Orobanche, Proceedings of the Third International Workshop on Orobanche and Related Striga Research*, Amsterdam, Netherlands, November 8-12,1993; *Royal Tropical Institute*; Amsterdam, Netherlands, 1994, pp. 539-544.

Bedi, J. Studies on *Fusarium oxysporum* for control of sunflower broomrape. *Haustorium*, 1992, 27, 1-2.

Bedi, J.S., Donchev, N. A technique for macroconidia production by *Fusarium oxysporum* f. sp. orthoceras, a biocontrol agent for *Orobanche. Plant Dis. Res.*, 1995, 10, 62-63.

Bedi, J.S., Donchev, N. Results on mycoherbicide control of sunflower broomrape (*Orobanche cumana* Wall.) under glasshouse and field conditions. In: Ransom J.K., Musselman L.J., Worsham A.D., Parker, C., (eds.), *Proceedings of the 5th International Symposium of Parasitic Weeds*, Nairobi, Kenya, 24-30 June 1991; Nairobi, Kenya.

Bedi, J.S., Sauerborn, J. A new technique for chlamydospore production by *Fusarium oxysporum orthoceras*, a mycoherbicidal agent for *Orobanche cumana. Plant Dis. Res.*, 1999, 14, 207-209.

Benharrat, H., Boulet, C., Théodet, C., Thalouarn, P. Virulence diversity among branched broomrape (*Orobanche ramosa* L.) populations in France. *Agron. Sustain. Dev.*, 2005, 25, 123–128.

Ben-Hod, G., Losner, D., Joel, D.M., Mayer, M. Pectin methylesterase in calli and germinating seeds of *Orobanche aegyptiaca. Phytochemistry*, 1993, 32, 1399-1402.

Bernhard, R.H., Jensen, J.E., Andreasen, C. Prediction of yield loss caused by *Orobanche* spp. in carrots and pea crops based on the soil seedbank. *Weed Res.*, 1998, 38, 191-197.

Boari, A., Vurro, M., Evidente, A., Andolfi, A., Fiore, M. Use of fungal metabolites for broomrape suicidal germination (Vurro). *Joint Working Groups and MC Meeting of COST Action 849, Broomrape Biology, Control and Management* 15-17 September 2005, Reading University, UK. Abstract Page 27.

Bouhatous, B., Jacquard, P. The effect of combinations of hosts on infection capacity of *Orobanche crenata* Forsk. In: Pieterse, A.H., Verklei, J.A.C., Borg, S.J. (eds.). *Biology and Management of Orobanche, Proceedings of the Third International Workshop on Orobanche and Related Striga Research*. Amesterdam, The Netherlands, Royal Tropical Institute, 1994, pp. 320-333

Boulet, C., Labrousse, P., Arnaud, M.C., Zehhar, N., Fer, A. Weed species present various responses to *Orobanche ramosa* L. attack. In: Fer, A., Thalouarn, P., Joel, D.M., Musselman, L.J., Parker, C., Verkleij, J.A.C., (eds.), *Proceedings of the 7th International Parasitic Weed Symposium,* Nantes, France, 2001, pp. 228-231.

Bouraoui, M., Abbes, Z., Abdi, N., Hemissi, I., Sifi, B. Evaluation of efficient *Rhizobium* isolates as biological control agents of *Orobanche foetida* Poir. parasitizing *Vicia faba* L. minor in Tunisia. *Bulg. J. Agric. Sci.*, 2012, 18, 557–564.

Bouraouia, M., Abbesc, Z., Rouissid, M., Abdi, N., Hemissia, I., Koukia, S., Sifi, B. Effect of rhizvara inoculation, N and P supply on *Orobanche foetida* parasitising faba bean (*Vicia faba* minor) under field conditions. *Biocontrol Sci. Technol.*, 2016, 26 (6),776-791. http://dx.doi.org/10.1080/09583157.2016.1157137 publishing, Ltd.

Brault, M., Betsou, F., Jeune, B., Tuquet, C., Salle, G. Variability of *Orobanche ramosa* populations in France as revealed by cross infestations and molecular markers. *Environ. Exp. Bot.*, 2007, 61, 272-280.

Braun, M., Burostaller, H., Eldin, A.M., Fadl, G. Approaches for the control of the parasitic weed *Orobanche ramosa* L. with special references to the use of glyphosate spraying and solar heating techniques on eggplant under the conditions of Sudan. *Acta Hortic.*, 1985, 158, 335-345.

Braun, M., Koch, W., Mussa, H. H., Stiefvater, M. Solarization for weed and pest control possibilities and limitations. In: Cavalloro, R., El Titi, A., (eds.), *Weed Control in Vegetable Production, Proceedings of a Meeting of the EC Experts' Group,* Stuttgart, 28-31 October 1986. Rotterdam/Brookfild: A. A, Balkema, 1988, pp. 169-178.

Braun, M., Koch, W., Stiefvater, M. Solarization possibilities and limitations demonstrated in trials in southern Germany and the Sudan. *Gesunde Pflanz.*, 1987, 39, 301-309.

Brenchley, W. E. *Weeds of Farm Land.* Longman, Green and Co., London. 1920, 239 p.

Brhane, T., Kiros-Meles, A., Chala, A., Asefa, D. Integrated management of *Orobanche crenata* in faba bean in Ethiopia. In: Nampala, M. P., Egeru, A., Tusiime, G., Osiru, M., Mensah, S., Adipala, E., (eds.), *Fifth African Higher Education Week and RUFORUM Biennial Conference 2016, "Linking agricultural universities with civil society, the private sector, governments and other stakeholders in support of agricultural development in Africa",* Cape Town, South Africa, 17-21 October 2016 pp.575-582.

Bronstejn, C.G., Kabulov, D.T. Perspectives of biological control of *Orobanche aegyptiaca*. *Trudy Samarkandskogo Gosudarstvennogo Universiteta Imeni Alisera Navoi. Samarkand,* 1961, 109, 73–89.

Brown, R., Greenwood, A.D., Johnson, A.W., Long, A.G., Lansdowne, A.R., Sunderland, N. The *Orobanche* germination factor. 3. Concentration of the factor by counter current distribution. *Biochem. J.*, 1952, 52, 557-574.

Buschmann, H., Giashi, K. G., Sauerborn, S. Induced resistance in sunflower against *Orobanche cumana*. In: Fer, A., Thalouarn, P., Joel, D.M., Musselman, L.J., Parker, C., Verkleij, J.A.C., (eds.), *Proceedings of the* 7th *International Parasitic Weed Symposium,* Nantes, France, 2001, Page 232.

Buschmann, H., Gonsior, G., Sauerborn, J. Pathogenicity of branched broomrape (*Orobanche ramosa*) populations on tobacco cultivars. *Plant Pathol.*, 2005, 54, 650–656.

Butler, L. G. Chemical Communication between the parasitic weed *Striga* and its crop host. A new dimension in allelochemistry. In: Inderjit, Dakshini, K.M.M., Einhellig, F.A., (eds.), *Allelopathy: Organisms, Processes, and Applications*, 1995, 158-168.

Cartia, G. Solar heating of the soil for the control of soil pests and perennial weeds. *Colture Protette*, 1985, 14, 7-42.

Castejon-Munoz, M., Romero-Munoz, F., Garcia-Torres, L. Effect of planting date on broomrape (*Orobanche cernua* Loefl.) infections in sunflower (*Helianthus annuus* L.). *Weed Res.*, 1993, 33, 171-176.

Castillejo, M.A., Fernández-Aparicio, M., Rubiales, D. Proteomic analysis by two-dimensional differential in gel electrophoresis (2D DIGE) of the early response of *Pisum sativum* to *Orobanche crenata*. *J. Exp. Bot.*, 2012, 63,107–119.

Chabrolin, C. Germination des graines et plantes-hotes de *Orobanche* de la feve (*Orobanche speciosa* D.C.). *Comptes Rendus de l'Academie des Sciences, Paris*, 1935, 200, 1974-1976.

Chalkov, Ch., Shabanov, D. Broomrape control and the possibilities provided by an introduced insect. *B'lgarski Tyutyun*, 1970, 15(1), 13–20.

Chari, M.S., Prabhu, S.R., Ramaprasad, G. Role of botanicals in pest management of tobacco in India. In: *Botanical Pesticides in Integrated Pest Management*. Rajahmundry, India, *Indian Society of Tobacco Science*, 1993, 128-133.

Charudattan, R. Biological control of weeds by means of plant pathogens: significance for integrated weed management in modern agro-ecology. *BioControl*, 2001, 46, 229–260

Chen, J., Xue, Q. H., McErlean, C. S. P., Zhi, J. H., Ma, Y. Q., Jiao, X. T., Zhang, M., Ye, X. X. Biocontrol potential of the antagonistic microorganism *Streptomyces enissocaesilis* against *Orobanche cumana*. *BioControl,* 2016, 61,781–791. DOI 10.1007/s10526-016-9738-z

Chittapur, B.M., Hunshal, C.S., Shenoy, H. Allelopathy in parasitic weed management: role of catch and trap crops. *Allelopathy J.*, 2001, 8, 147-160.

Christeva, T., Naumova, S. Stimulation of broomrape seed germination by soil microorganisms. In: Wegmann, K., Musselman, L.J., Joel, D.M., (eds.), *Current Problems of Orobanche Researches. Proceedings of the Fourth International Workshopon* Orobanche, Albena, 23-26, September, Bulgaria, 1998, 73-77.

Cimmino, A., Fernández-Aparicio, M., Avolio, F., Yoneyama, K., Rubiales, D., Evidente, A. Ryecyanatines A and B and ryecarbonitrilines A and B, substituted cyanatophenol, cyanatobenzo (1,3) dioxole, and benzo (1,3) dioxolecarbonitriles from rye (*Secale cereale* L.) root exudates: Novel metabolites with allelopathic activity on *Orobanche* seed germination and radicle growth. *Phytochemistry*, 2015, 109, 57–65.

Cohen, B., Amsellem, Z., Lev-Yadun, S., Gresse, L, J. Infection of tubercles of the parasitic weed *Orobanche aegyptiaca* by mycoherbicidal *Fusarium* species. *Ann. Bot.*, 2002, 90, 567–578.

Cubero, J. I., Hernandez, L. Breeding faba bean (*Vicia faba* L.) for resistance to *Orobanche crenata* Forsk. In: Cubero, J.I., Saxena, M.C., (eds.), *Present Status and Future Prospects of Faba Bean Production and Improvement in the Mediterranean Countries, Options Mediterraneennes, Serie A: Seminares Mediterraneennes, 10.* Zaragoza, Spain. CHEAM/ECC/ECARDA. 1991, pp. 51-57.

Cubero, J. I., Moreno, M. T., Rubiales, D., Sillero, J., (eds.). *Resistance to Orobanche: The State of Art.* Dirección General de Investigación y Formación Agraria, Sevilla, Spain. 1999.

Cubero, J.I. Breeding for resistance to *Orobanche* species: a review. In: Wegmann K., Musselman L.J., (eds.), *Progress in* Orobanche *Research. Proceedings, International Workshop on* Orobanche *Research,* Obermarchtal, 1989. Eberhard-Karls-Universitat, Tubingen, 1991, pp. 257-277.

Cubero, J.I. Resistance to *Orobanche crenata* Forsk. in *Vicia faba* L. In: Anonymous., *Proccedings of the European Weed Research Council Symposium, Parasitic Weeds*, The European Weed Research Council, Parasitic Weeds Research Group, Wageningen, The Netherlands. Printed by Malta University Press. 1973, pp. 205-217.

Cubero, J.I., Pieterse, A. H., Khalil, S.A., Sauerborn, J. Screening techniques and source of resistance to parasitic angiosperms. In: Muehlbauer, F. J. , Kaiser, W. J., (eds.), *Expanding the Production and Use of Cool Season Food Legumes. Proceedings of the Second International Food Legume Research Conference on Pea, Lentil, Faba bean, Chickpea and Grasspea,* April12-16, 1992, Cairo, Egypt.

Cubero, J.I., Pieterse, A.H., Khalil, S.A., Sauerborn, J. Screening techniques and sources of resistance to parasitic angiosperms. *Euphytica*, 1994, 73, 51–58.

Cvejic, S., Dedic, B., Jocic, S., Miladinovic, D., Miklic, V. Broomrape resistance in newly developed sunflower inbred lines. In: *Proceedings of the 18th International Sunflower Conference*; 2012 Feb 27–Mar 1; Mar del Plata, Argentina, Paris: *The International Sunflower Association (ISA) and the Argentine Sunflower Association (ASAGIR)*, Argentina,, 2012, 1037–1042.

Dalela, G.G., Mathur, R.L. Resistance of varieties of eggplant, tomato and tobacco to broomrape (*Orobanche cernua* Loefl.). *Pest Article and News Summaries* (*PANS*), 1971, 17, 482-483.

Daniel, J., Chaudhuri, S., Plakhine, D., Ziadna, H., Steffens, J. Dehydrocostus lactone is exuded from sunflower roots and stimulates germination of the root parasite *Orobanche cumana. Phytochemistry*, 2011, 72, 624–634.

Danko, J. Effectiveness of preparations based on the hydrazide of maleic acid against *Orobanche ramosa* L. in tobacco. *Polnohospodarstvo*., 1993a, 37, 923-931.

Danko, J. Use of herbicides for control of Broomrape (*Orobanche ramosa* L.) on tobacco. *Ochrana Rostlin*., 1993b, 29, 303-307.

Darbinyan, G.A., Saakyan, R.I., Gabrielyan, Zh. P. Testing of triethanolamine salt of maleic hydrazide (MH-T) against *Orobanche* in tobacco plantations. In Tezisy 8-1 sessi Zakavkaziskogo soveta po koordinatsii nauchno-issledovatel'skikh rabot po zashchite rastenii. *Erevan, Armenian SSR*., 1977, 473-474.

Darvishzadeh, R. Genetic variability, structure analysis, and association mapping of resistance of resistance to broomrape (*Orobanche aegyptiaca* Pers.) in tobacco. *J Agric Sci Technol*., 2016, 18, 1407–1418.

Demirkan, H., Nemli, Y. Effects of some fertilizers on *Orobanche ramosa* L. on tomato. In: *Biology and Management of Orobanche.* In: Pieterse, A.H.; Verkleij, J.A.C., ter Borg, S.J., (eds.), *Proceedings of the Third International Workshop on Orobanche and Related Striga Research,* Amsterdam, Netherlands, November 8-12 , 1993. Royal Tropical Institute, 1994, Amsterdam, Netherlands.

Denev, I., Pereira, A., Verstappen, F., Bouwmeester, H. Biosynthesis of *Orobanche* germination stimulants. In: Fer, A., Thalouarn, P., Joel, D.M., Musselman, L.J., Parker, C., Verkleij, J.A.C., (eds.), *Proceedings of the* 7th *International Parasitic Weed Symposium,* Nantes, France. 2001, 110-113.

Dhanapal, G.N. *Management of Broomrape* (Orobanche cernua) *in Tobacco* (Nicotiana tabacum). WAU Dissertation no. 1996, 2135.

Dhanapal, G.N., Mallory-Smith, P.C., Ter Borg, S.J. Interactions between nodding broomrape and bidi tobacco in India. In: Fer, A., Thalouarn, P., Joel, D.M., Musselman, L.J., Parker, C., Verkleij, J.A.C., (eds.), *Proceedings of the 7th International Parasitic Weed Symposium,* Nantes, France, 2001, p. 42.

Dhanapal, G.N., Struik, P.C. Broomrape (*Orobanche cernua*) control before attachment to host through chemically or biologically manipulating and germination. *Neth. J. Agric. Sci.*, 1996, 44, 279-291.

Dhanapal, G.N., Struik, P.C. Natural plant oils: Do they kill broomrape spikes? In: *Current Problems of Orobanche Research.* In: Wegmann, K., Musselman, L.J., Joel, D.M., (eds.), *Proceedings of the Fourth International Workshop on* Orobanche, September 23-26, Bulgaria, 1998, 177-180.

Dhanapal, G.N., Struik, P.C., ter Borg, S.J. Effect of natural stimulants with and without GR24 on broomrape germination. In: Wegmann, K., Musselman, L.J., Joel, D.M., (eds.), *Current Problems of* Orobanche *Research. Proceedings of the Fourth International Workshop on* Orobanche, Albena, 23-26, September, Bulgaria, 1998, pp. 51-57.

Dhanapal, G.N., Struik, P.C., Udayakumar, M., Timmermans, P.C.J.M. Management of broomrape (*Orobanche* spp.)- a review. *J. Agron. Crop Sci.*, 1996, 176, 335-359.

Diaz, J., Norambuena, H. Parasitism and phenology of *Orobanche ramosa* L. on tomato. In: Fer, A., Thalouarn, P., Joel, D.M., Musselman, L.J., Parker, C., Verkleij, J.A.C., (eds.), *Proceedings of the 7th International Parasitic Weed Symposium,* Nantes, France, 2001, 47.

Dinesha, M.S., Dhanapal, G.N., Prabhudev Dhumgond, N.S., Vignesh, V., Madhukumar, K Raghavendra. Efficiency and economics of broomrape (*Orobanche cernua* Loefl.) control with herbicides in infested tomato (*Lycopersicon esculentum* Mill.) field. *Plant Arch.*, 2012, 12(2), 833-836.

Disciglio, G., Carlucci, A., Tarantino, A., Giuliani, M.M., Gagliardi, A., Frabboni, L., Libutti, A., Raimondo, M.L., Lops, F., Gatta, G. Effect of olive-mill wastewater application, organo-mineral fertilization, and transplanting date on the control of *Phelipanche ramosa* in open-field processing tomato crops. *Agronomy*, 2018, 8, 92. Doi: 10.3390/agronomy8060092, 13pp.

Disciglio, G., Lops, F., Carlucci, A., Gatta, G., Tarantino, A., Frabboni, L., Carriero, F., Tarantino, E. Effects of different methods to control the parasitic weed *Phelipanche ramosa* (L.) Pomel in processing tomato crops. *Ital. J. Agron.*, 2016, 11, 39–46.

Dominguez, J. R-41, a sunflower restorer inbred line, carrying two genes for resistance against a highly virulent Spanish population of *Orobanche cernua. Plant Breed.*, 1996, 115, 203-204.

Dominguez, J., Melero- Vara, J.M., Ruso, J., Miller, J.F., Fernandez- matine, J. M. Screening forresistance to broomrape (*Orobanche cernua*) in cultivated sunflower. USDA animal and planthealth inspection service, USDA Plant Protection and Quarantine, Georgia. Department of agriculture, Georgia cooperative extension service, and gergia department of transportation, March 1993. Small broomrape in Baker country, Georgia.

Dominguez, J., Melero-Vara, J. M., Ruso, J., Miller, J.F., Fernandez-Martinez, J.M. Screening for resistance to broomrape (*Orobanche cernua*) in cultivated sunflower. *Plant Breed.*, 1996, 115, 201-202.

Dong, S. Q., Ma, Y. Q., Wu, H., Shui, J. F., Hao, Z. Q. Stimulatory effects of wheat (*Triticum aestivum* L.) on seed germination of *Orobanche minor* Sm. *Allelopathy J.*, 2012,30 (2), 247-258.

Dongola G. M., Mohamed, A.K., Ahmed, M.K. Effect of some control methods for broomrape. Haalouk (*Orobanche ramosa*) in the Sudan. In: *Orobanche in the Central Sudan.* The FAO/ARC IPM Project GCP/SUS/025/NET, P.O. Box 196. Wad Medani, Sudan. February, 2000. Khartoum, Sudan.

Dongola, G. M. Effect of crop sequence on *Orobanche ramosa* management in tomato crop. *J. Sc. Tech.* 2006, 7(1), 1-8.

Dor, E., Alperin, B., Wininger, S., Ben-Dor, B., Somvanshi, V.S., Koltai, H., Kapulnik, Y., Hershenhorn, J. Characterization of a novel tomato mutant resistant to the weedy parasites *Orobanche* and *Phelipanche* spp. *Euphytica*, 2010, 171, 371–380 DOI 10.1007/s10681-009-0041-2

Dor, E., Smirnov, E., Galili, S., Guy, A.; Hershenhorn, J. Characterization of the novel tomato mutant HRT, resistant to acetolactate synthase–inhibiting herbicides. *Weed Sci.*, 2016, 64, 348–360.

Dor, E., Yoneyama, K., Wininger, S., Kapulnik, Y., Yoneyama, K., Koltai, H., Xie, X., Hershenhorn, J. Strigolactone deficiency confers resistance in tomato line *SL-ORT1* to the parasitic weeds *Phelipanche* and *Orobanche* spp. *Phytopathology*, 2011, 101, 213-222.

Dorr, I., Kollmann, R. Structural features of parasitism of *Orobanche* II. The differentiation of assimilate conducting elements within the haustorium. *Protoplasma*, 1975, 83, 185-199.

Dorr, I., Kollmann, R. Structural features of parasitism of *Orobanche*. III. The differentiation of xylem connection of *O. crenata*. *Protoplasma*, 1976, 89, 235-239.

Drennan, D.S.H., Mohamed-Ahmed, A.G. Some effects of temperature on germination of *Orobanche ramosa* L. In: *Ixeme Colloque International sur la Biologie des Mauvaises Herbes*, 16-18 September 1992, Dijon, France. Paris, France; ANPP.COLOMA/EWRS, pp. 117-120.

Dubey, N.K., Eizenberg, H., Leibman, D.,Wolf, D., Edelstein, M., Abu-Nassar, J., Marzouk, S., Gal-On, A., Aly, R. Enhanced host-parasite resistance based on down-regulation of *Phelipanche aegyptiaca* target genes is likely by mobile small RNA. *Front. Plant Sci.* 2017, 8, 1574.

Edwards, W.G. H. *Orobanche* research at the Royal University of Malta. *Pest Article and News Summaries* (*PANS*), 1972, 18, 19-22.

Eizenberg, H., Plakhine, D., Dor, E., Hershenhorn, J., Kleifeld, Y., Rubin, B. Phytotoxic root extract from resistant sunflower (*Helianthus annuus* L. cv. Ambar) inhibits *Orobanche cumana* development. In: Fer, A., Thalouarn, P., Joel, D.M., Musselman, L.J., Parker, C., Verkleij, J.A.C., (eds.), *Proceedings of the 7th International Parasitic Weed Symposium,* Nantes, France., 2001, 190-191.

El-Hattab, A.H., Abdalla, M.M. F., Metwally, A.A., Darwish, D.S. Effect of glyphosate on *Orobanche* parasitism in a collection of faba beans. *Research Bulletin.* Faculty of Agriculture, Ain Shams University, 1982, 2101, 20pp.

Elia, Y. I., Al-Sahaf, N. A., Al-Khesraji, T. O. Preliminary studies on *Orobanche aegyptiaca* Pers. in tomato.) *Iraqi J. Agric. Sci. "Zanco'*, 1987, 5, 233-244.

El-Kassas, R., Karam El-Din Z., Beale, M.H., Ward, J.L., Strange, R.N. Bioassay-led isolation of *Myrothecium verrucaria* and verrucarin A as germination inhibitors of *Orobanche crenata. Weed Res.*, 2005, 45, 212–219.

Eltayeb, D.M., Thalouran, P., Fer, A., Babiker, A.G.T., Mohamed, Y.F., Hamdoun, A.M., Dafalla, G.A. Effect of broomrape (*O. ramosa*) seed placement on parasitism of tomato (*L. esculentum* Mill.) plants. In: Americanos, P.G., Abu-Irmaileh, B. E., Saghir, A.R., (eds.), *Proceedings of the First Near East Conference on Improved Weed Management*, February, 2000. Khartoum, Sudan, 194 -198.

Elzein, A.E.M. Development of a granular mycoherbicidal formulation of *Fusarium oxysporum* 'Foxy 2' for the biological control of *Striga hermonthica.* In: Kroschel, J., (ed.), *Tropical Agriculture - Advances in Crop Research.* 2003, 12 (2). Margraf Verlag, Weikersheim, Germany.

Elzein, A.E.M., Kroschel, J., Assefa, A., Masresha, F. Preliminary evaluation of *Phytomyza orobanchia* (Diptera: Agromyzidae) as a controller of *Orobanche* spp. in Ethiopia. *Sinet, Ethiopian Journal of Science*, 1999, 22, 271-282.

Eplee, R.E. *Orobanche ramosa* in the United States. In: Parker, C., Musselman, L.J., Polhill, R. M., Wilson, A.K., (eds.), *Proceedings, 3rd International Symposium on Parasitic Weeds*, International Center for Agricultural Research in the Dry Areas (ICARDA), Aleppo, Syria. 1984, pp. 40-42.

Eplee, R.E., Norris, R., Merritt, J. Effect of nitrogen on parasitic weed seed. In *Biology and Management of* Orobanche. Pieterse A. H., Verkleij J.A.C., ter Borg, S.J., (eds.), *Proceedings of the Third International Workshop on Orobanche and Related Striga Research,* November 8-12, 1993, Amsterdam, Netherlands. 1994.

Ernest, W.H.O. Mineral nutrition of *Nicotiana tabacum* cv. Bursana during infection by *Orobanche ramosa*. In: ter Borg S.J., (ed.), *Proceedings, Workshop on Biology and Control of* Orobanche, LH/VPO, Wageningen, 1986 pp. 80-85.

Evidente, A., Andolfi, A., Fiore, M., Boari, A., Vurro, M. Stimulation of *Orobanche ramosa* seed germination by fusicoccin derivatives: A structure–activity relationship study. *Phytochemistry.* 2006, 67, 19–26.

Evidente, A., Cimmino, A., Fernández-Aparicio, M., Andolfi, A., Rubiales, D., Motta, A. Polyphenols, including the new pea polyphenols A–C, from root exudates stimulate *Orobanche foetida* seed germination. *J. Agric. Food Chem.* 2010, 58, 2902–2907.doi:10.1021/jf904247k.

Evidente, A., Cimmino, A., Fernández-Aparicio, M., Rubiales, D., Andolfi, A., Melck, D. Soya sapogenol Band trans-22-dehydrocamposterol from common vetch (*Vicia sativa* L.) root exudates stimulate broomrape seed germination. *Pest Manag. Sci.* 2011, 67, 1015–1022.doi:10.1002/ ps.2153.

Evidente, A., Fernández-Aparicio, M., Andolfi, A., Rubiales, D., Motta, A. Trigoxazonane, a monosubstituted trioxazonane from *Trigonella foenum-graecum* root exudate, inhibits *Orobanche crenata* seed germination. *Phytochemistry*, 2007, 68, 2487–2492.

Evidente, A., Fernández-Aparicio, M., Cimmino, A., Rubiales, D., Andolfi, A., Motta, A. Peagoland peagoldione, two new strigolactone like metabolites isolated from pea root exudates. *Tetrahedron Lett.,* 2009, 50, 6955–6958. doi: 10.1016/j.tetlet.2009.09.142.

FAO. *Progress on Farmer Training in Parasitic Weed Management*. Rome-Italy. 2008. Fer, A., Thalouarn, P. *Orobanche*: a threat to our crops. *Phytoma.*, 1997, 50, 34-36, 38-40.

Ferna´ndez-Aparicio, M., Sillero, J. C., Rubiales, D. Resistance to broomrape in wild lentils (*Lens* spp.). *Plant Breed.*, 2009a, 128, 266-270.

Ferna´ndez-Aparicio, M., Sillero, J.C., Rubiales, D. Resistance to broomrape species (*Orobanche* spp.) in common vetch (*Vicia sativa* L.). *Crop Prot.*, 2009b, 28, 7-12.

Fernández-Aparicio, M., Andolfi, A., Evidente, A., Pérez-de-Luque, A., Rubiales, D. Fenugreek root exudates show species specific stimulation of *Orobanche* seed germination. *Weed Res.*, 2008a, 48, 163–168.

Fernández-Aparicio, M., Cimmino, A., Evidente, A., Rubiales, D. Inhibition of *Orobanche crenata* seed germination and radicle growth by allelochemicals identified in cereals. *J. Agric. Food Chem.*, 2013, 61, 9797–9803.

Fernández-Aparicio, M., Emeran, A.A., Rubiales, D. Control of *Orobanche crenata* in legumes intercropped with fenugreek (*Trigonella foenum-graecum*). *Crop Prot.*, 2008b, 27, 653-659.

Fernández-Aparicio, M., Emeran, A.A., Rubiales, D. Inter-cropping with berseem clover (*Trifolium alexandrinum*) reduces infection by *Orobanche crenata* in legumes. *Crop Prot.*, 2010a, 29, 867–871.

Fernández-Aparicio, M., Garcıa-Garrido, J.M., Ocampo, J.A., Rubiales, D. Colonisation of field pea roots by arbuscular mycorrhizal fungi reduces *Orobanche* and *Phelipanche* species seed germination. *Weed Res*., 2010b, 50, 262–268.
Fernández-Aparicio, M., Moral, A., Kharrat, M., Rubiales, D. Resistance against broomrapes (*Orobanche* and *Phelipanche* spp.) in faba bean (*Vicia faba*) based in low induction of broomrape seed germination. *Euphytica*, 2012, 186, 897–905.
Fernández-Aparicio, M., Reboud, X., Gibot-Leclerc, S. Broomrape Weeds. Underground mechanisms of parasitism and associated strategies for their control: A Review. *Front. Plant Sci*., 2016, 7, 135. doi: 10.3389/fpls.2016.00135
Fernández-Aparicio, M., Rubiales, D. Characterization of resistance to crenate broomrape (*Orobanche crenata* Forsk.) in *Lathyrus cicera* L. *Euphytica*, 2010, 173, 77–84.
Fernández-Aparicio, M., Sillero, J.C., Rubiales, D. Intercropping with cereals reduces infection by *Orobanche crenata* in legumes. *Crop Prot*., 2007, 26, 1166–1172.
Fernández-Aparicio, M., Yoneyama, K., Rubiales, D. The role of strigolactones in host specificity of *Orobanche* and *Phelipanche* seed germination. *Seed Sci Res*., 2011, 21, 55–61.
Fire, A., Xu, S., Montgomery, M.K., Kostas, S.A., Driver, S.E., Mello, C.C. Potent and specific genetic interference by double stranded RNA in Caenorhabditis elegans. *Nature*, 1998, 391, 806–811.
Foy, C.L., Jacobsohn, R. Screening tomato lines for glyphosate tolerance. In: *Proceedings, Southern Weed Science Society, 36th Annual Meeting*. Jan. 18-20, 1983, Biloxi, Mississippi, *Southern Weed Science Society*. P.165.
Foy, C.L., Jacobsohn, R., Jain, R. Screening of *Lycopersicon* spp. for glyphosate and/or *Orobanche aegyptiaca* Pers. Resistance. *Weed Res*., 1988, 28, 383-291.
Foy, C.L., Jain, R., Jacobsohn, R. Recent approaches for chemical control of broomrape (*Orobanche* spp.). *Review of Weed Sci*., 1989, 4, 123-152.
Galun, E. RNA silencing. Singapore: World Scientific; Galun, E. RNA silencing in plants. *In Vitro Cell. Dev. Biol., Plant*., 2005, 41,113–123.
Garcia-Torres, L. Progress in *Orobanche* control, an overview. In: Pieterse A.H.; Verkleij J.A.C., ter Borg, S.J. (eds.), *Biology and Management of Orobanche. Proceedings of the Third International Workshop on Orobanche and Related Striga Research,* November 8-12, 1993, Amsterdam, Netherlands.
Garcia-Torres, L., Castegon-Munoz, M., Lopez-Granados, F., Jurado-Exposito, M. Imazapyr applied postemergence in sunflower (*Helianthus annuus*) for broomrape (*Orobanche cernua*) control. *Weed Technol*., 1995, 9, 819-824.
Garcia-Torres, L., Lopez-Granados, F. Control of broomrape (*Orobanche crenata* Forsk.) in broad bean (*Vicia faba* L.) with imidazolinones and other herbicides. *Weed Res*., 1991a, 31, 227-235.
Garcia-Torres, L., Lopez-Granados, F. Progress of herbicide control of broomrape (*Orobanche* spp.) in legumes and sunflower (*Helianthus annuus* L.). In: Ransom, J.K., Musselman, L.J., Worsham, A.D., Parker, C., (eds.), *Proceedings of the 5th International Symposium of Parasitic Weeds,* June 24-30, 1991b Nairobi, Kenya.
Garcia-Torres, L., Lopez-Granados, F., Jurado-Exposito, M., Diaz-Sanchez, J. The present state of *Orobanche* spp. infestations in Andalusia and the prospects for its management. In: Maillet, J., (ed.), *6th EWRS Mediterranean Symposium*, Montpellier, France, 1998. pp. 141-145.
Gauthier, M., Veroneci, C., El-Halmouch, Y., Leflon, M., Jestin, C., Labalette, F., Simier, P. Characterisation of resistance to branched broomrape, *Phelipanche ramosa*, in winter oilseed rape. *Crop Prot*., 2012, 42, 56–63.
Gayed, S. K. Dodder in tobacco seedbeds in Ontario and its control. *Can. J. PI. Sci.,* 1986, 66,421-423.

Geipert, S. Entwicklung integrierter Verfahren zur Kontrolle von *Orobanche crenata* Forsk. in Marokko. *Internal Report, Supra-regional GTZ-project "Ecology and Management of Parasitic Weeds*", University of Hohenheim, Germany, unpublished. 1993.

Geipert, S. *Potentials and Constraints for the Control of Orobanche crenata Forsk. In Faba Beans* (*Vicia faba* L.) *in Morocco*. Stuttgart, Germany; Institute fur Pflanzenproduktion in den Tropen and Subtropen, Universitat Hohenheim, 1997,144 PP.

Georgieva, I.G., Bozoukov, H.G. Study on the properties of some pesticides to stimulate germination of the broomrape (*Orobanche* spp.) seeds. *Bulg. J. Agric. Sci.*, 1995, 1(1), 13-17.

Gevezova, M., Dekalska, T., Stoyanov, K., Hristeva, T., Kostov, K., Batchvarova, R. and Denev, I. Recent advances in broomrapes research. *J. BioSci. Biotech.,* 2012, 1(2), 91-105.

Ghannam, I., Al-Masri, M., Barakat, R. The Effect of herbicides on the Egyptian broomrape (*Orobanche aegyptiaca*) in tomato fields. *Am. J. Plant Sci.*, 2012, 3, 346–352.

Gharib, M.S. Biological and economical aspects of the broomrapes (*Orobanche* spp.) in Northern Iraq. In: Anonymous., *Proceedings, Symposium on Parasitic Weeds, Malta,* 1973. *European Weed Research Council*, Wageningen, 1973 pp., 44-47.

Ghosheh, H.Z., Al-Tamimi, E., Hameed, K.M. Effect of olive jift and sublethal glyphosate applications on faba beans (*Vicia faba*). *Acta Agron. Hung.* 2006, 54, 61–68.

Ghosheh, H.Z., Hameed, K.M., Turk, M.A., Al-Jamali, A.F. Olive (*Olea europaea*) jift suppresses broomrape (*Orobanche* spp.) infections in faba bean (*Vicia faba*), pea (*Pisum sativum*) and tomato (*Lycopersicon esculentum*). *Weed Technol.*, 1999, 13, 457-460.

Ghotbi, M., Rouhi, H.R., Amini Dehaghi, M., Ghotbi, M., Moghaddam Khamseh, A.R.,Wahsha, M. Mitigate *Phelipanche aegyptiaca* Pers. infestation considering natural environment conservation. *International Journal of AgriScience*, 2012, 2(1), 62-77. Available at: www.inacj.com.

Gil, J.; Martin, L.M., Cubero, J.I. Genetics of resistance in *Vicia sativa* L. to *Orobanche crenata* Forsk. *Plant Breed.*, 1987, 99, 134-143.

Ginman, E., Prider, J., Matthews, J., Virtue, J., Watling, J. Sheep as vectors for branched broomrape (*Orobanche ramosa* subsp. *mutelii* (F.W. Schultz) Cout.) seed dispersal. *Weed Biol. Manag*., 2015, 15, 61–69.

Giray, H., Nemli, Y. Investigations on the morphological characters, brief life history and effectiveness of the natural enemy of *Orobanche, Phytomyza orobanchia* Kalt. (Diptera, Agromyzidae) in Izmir Province.) *Turkiye Bitki Koruma Dergisi*, 1983, 7(3), 183–192.

Gold, A.H., Sagen, J.E., Wilhelm, S. California soils suppressive to branched broomrape (*Orobanche ramosa* L.). In: Musselman, L.J., Worsham, A.D., Eplee, R.E. (eds.), *Supplement to Proceedings of the* 2nd *International Symposium on Parasitic Weeds.* North Carolina State University, Raleigh, USA. 1979, p. 1.

Goldwasser, Y., Eizenberg, H., Golan, S., Kleifeld, Y. Control of *Orobanche crenata* and *Orobanche aegyptiaca* in parsley. *Crop Prot.*, 2003, 22, 295–305

Goldwasser, Y., Eizenberg, H., Hershenhorn, J., Plakhine, D., Blumenfeld, T., Buxbaum, H., Golan, S., Kleifeld, Y. Control of *Orobanche aegyptiaca* and *O. ramosa* in potato. *Crop Prot.*, 2001, 20, 403-410.

Goldwasser, Y., Kleifeld, Y. Tolerance of parsley varieties to *Orobanche. Crop Prot.*, 2002, 21, 1101–1107.

Goldwasser, Y., Kleifeld, Y., Golan S., Bargutti, A., Rubin, B. Metham sodium's dissipation from the soil and its effect on the control of Egyptian broomrape (*Orobanche aegyptiaca* Pers.). In: *Biology and Management of Orobanche*. In: Pieterse, A.H., Verkleij, J.A.C., ter Borg, S.J., (eds.), *Proceedings of the Third International Workshop on Orobanche and Related Striga Research,* November 8-12,1993. Amsterdam, Netherlands.

Goldwasser, Y., Kleifeld, Y., Golan, S., Bargutti, A., Rubin, B. Dissipation of metham-sodium from soil and its effect on the control of *Orobanche aegyptiaca*. *Weed Res.*, 1995, 35, 445-452.

Goldwasser, Y., Kleifeld, Y., Joel, D.M., Plakhine, D. Variations in vetch (*Vicia* spp.) response to *Orobanche aegyptiaca* Pers. In: Moreno, M.T., Cubero, J.I., Berner, D., Joel, D., Musselmman, L.J., Parker, C., (eds.), *Advances in Parasitic Plant Research, Proceedings of the Sixth International Parasitic Weed Symposium.* April 17-18, 1996. Cordoba, Spain, 1996 pp. 615-624.

Goldwasser, Y., Kleifeld, Y., Plakhine, D., Robin, B.Variation in vetch (*Vicia* spp.) response to *Orobanche aegyptiaca*. *Weed Sci.*, 1997, 45, 756-762.

Goldwasser, Y., Plakhine, D., Yoder, J.I. *Arabidopsis thaliana* susceptibility to *Orobanche* spp. *Weed Sci.*, 2000, 48, 342-346.

Goldwasser, Y., Rodenburg, J. Integrated agronomic management of parasitic weed seed banks, 2014. DOI:10.1007/978-3-642-38146-1_22.

Grandos, F.L., Torres, L.G. Seed bank and other demographic parameters of broom rape (*Orobanche crenata* Forsk.) populations in fababean (*Vicia faba* L.). *Weed Res.*, 1993, 33, 319-327.

Gressel, J. *Molecular Biology of Weed Control.* Taylor & Francis, London. 2002.

Gressel, J., Hana, A., Head, G., Marasas, W., Obilana, B., Ochanda, J., Souissi, T., Tzotzos, G. Major heretofore intractable biotic constraints to African food security that may be amenable to novel biotechnological solutions. *Crop Prot.*, 2004, 23, 661–689.

Guitierrez-Mellado, M.C., Edwards, R., Tena, M., Cabello, F., Serghini, K., Jorrin, J. The production of coumarin phytoalexins in different plant organs of sunflower (*Helianthus annuus* L.). *J. Plant Physiol.*, 1995, 149, 261-266.

Guti'errez, N., Palomino, C., Satovic, Z., Ruiz-Rodríguez, M. D., Vitale, S., Gutiérrez, M. V., Rubiales, D., Kharrat, M., Amri, M., Emeran, A. A., Cubero, J. I., Atienza, S. G., Torres, A. M., Avila, C. M. QTLs for *Orobanche* spp. resistance in faba bean: identification and validation across different environments. *Mol. Breeding*, 2013, 32, 909–922.

Habimana, S., Nduwumuremyi, A., Chinama R. J.D. Management of *Orobanche* in field crops-a review. *J. Soil Sci. Plant Nut.*, 2014, 14(1), 43-62.

Haidar, M.A., Bibi, W., Abdel-Khalek, N. Effect of wheat and barley residues on branched broomrape (*Orobanche ramosa*) growth and development in potatoes. In: *Brighton Crop Protection Conference, Weeds. Proceedings of an International Conference,* Brighton, UK, November 20-23, 1995. Farnham, UK, *British Crop Protection Council,* 3, 871-876.

Haidar, M.A., Iskandarani, N., Sidahmed, M.M., Darwish, R. Susceptibility of *Orobanche ramosa* and potato tolerance to rimsulfuron. *Crop Prot.*, 2005b, 24, 7–13.

Haidar, M.A., Sidahmed, M.M. Elemental sulphur and chicken manure for the control of branched broomrape (*Orobanche ramosa*). *Crop Prot.*, 2006, 25, 47–51.

Haidar, M.A., Sidahmed, M.M. Soil solarization and chicken manure for the control of *Orobanche crenata* and other weeds in Lebanon. *Crop Prot.*, 2000, 19, 169-173.

Haidar, M.A., Sidahmed, M.M., Darwish, R., Lafta, A. Selective control of *Orobanche ramosa* in potato with rimsulfuron and sub-lethal doses of glyphosate. *Crop Prot.*, 2005a, 24, 743–747.

Hamid, A.E., M.M., Shebl, S. M., Ghalwash, A.M. New herbecide for control broomrape and other annual weeds in peas (*Pisum sativum*) at Kafr El-Sheikh. In: Americanos, P.G., Abu-Irmaileh, B.A., Saghir, A.R., (eds.), *Proceedings of the First Near East Conference on Improved Weed Management*, February, 2000. Khartoum, Sudan, 2000 pp. 245- 250.

Hammad, S., El-Arosi, H., Al-Menoufi, O. *Phytomyza orobanchia* Kalt. Feeding on *Orobanche crenata* Forsk. in Egypt (Diptera: Agromyzidae). *Bulletin de la Société Entomologique d'Egypte*, 1967, 51, 141–144.

Harb, A. *The Role of Indol-3-Acetic Acid in Host/Orobanche Interaction.* M.Sc. Thesis, Jordan University of Science and Technology, Irbid, Jordan. 2002, 104 PP.

Harloff, H.J., Wegmann, K. Mannitol pathway in *Orobanche.* In: Weber, H.C., Forstreuter, W., (eds.), *Proceedings of the Fourth International Symposium of Parasitic Flowering Plants,* Marburg, Germany, Philipps-Universitat, Murburg, 1987, pp 295–308.

Hassan E.A. Broomrape species in Egypt, a recent survey in relation to geographical distribution. In: Maillet, J. (ed.), *6th EWRS Mediterranean Symposium,* Montpellier, France, 1998, 155.

Hassan, E.A., Farrag, A.A. Host-parasite relationship. The influence of boron application at earlier stages of infection on growth behaviour of broad bean (*Vicia faba*) and *Orobanche* plants. Research Bulletin, Faculty of Agriculture, Ain Shams University No. 2017, 1982, 19 pp.

Hassanein, E.E., Fayad, Y. H., Shalaby, F.F., Kkolosy, A.S. Natural role of *Phytomyza orobanchia* Kalt., a beneficial fly against the parasitic weeds *Orobanche* spp. infesting legumes and carrots in Egypt. *Ann. Agric. Sci.* (*Cairo*), 1998, 43, 201-206.

Hauck, C, Müller, S, Schildknecht, H. A germination stimulant for parasitic flowering plants from *Sorghum bicolor*, a genuine host plant. *J. Plant Physiol.*, 1992, 139, 474–478.

Hemissi, I., Mabrouk, Y., Abdi, N., Bouraoui, M., Saidi, M., Sifi, B. Growth promotion and protection against *Orobanche foetida* of chickpea (*Cicer aerietinum*) by two *Rhizobium* strains under greenhouse conditions. *Afr. J. Biotechnol.*, 2013, 12(12), 1371–1377.

Hershenhorn, J., Goldwasser, Y., Plakhine, D., Herzlinger, G., Golan, S., Russo, R., Kleifeld, Y. Role of pepper (*Capsicum annuum*) as a trap and catch crop for control of *Orobanche aegyptiaca* and *O. cernua. Weed Sci.*, 1996, 44, 948-951.

Hershenhorn, J., Goldwasser, Y., Plankhine, D., Ali, R., Blumenfeld, T., Bucsbaum, H., Herzlinger, G., Golan, S., Chelf, T., Eizenberg, H., Dor, E., Kleifeld, Y. *Orobanche aegyptiaca* control in tomato fields with sulfonylurea herbicides. *Weed Res.*, 1998, 38, 343-349.

Herzlinger, G., Kleifeld, Y. Effect of the herbicide propyzamide on broomrape damage to specific host plants. *Phytoparasitica*, 1985, 13, 242 (Abstract).

Hodosy, S. Biological control of broomrape, *Orobanche ramosa*, a tomato parasite. In: *Occurrence and Adaptability of Fusarium species to Control Broomrape in Hungary. Zoldsegtermesztesi Kutato lntzei Bulletinje*, 1981, 14, 21-29.

Horváth, Z. Investigations on *Phytomyza orobanchia* Kalt. (Diptera: Agromyzidae), a possible biocontrol agent of *Orobanche* spp. (Orobanchaceae) in Hungary. In: Weber, H., Forstreuter, W., (eds.), *Proceedings of the 4th International Symposium on Parasitic Flowering Plants.* Marburg, BRD. 1987, pp. 403–410.

Horváth, Z. *Phytomyza orobanchia* Kalt. /Agromyzidae, szádor – aknázólégy elõfordulása napraforgó szádor növenyekben. *Növényvédelem*, 1981, 17(10–11), 436–442.

Horváth, Z. The role of the fly *Phytomyza orobanchia* Kalt. (Diptera: Agromyzidae) in reducing parasitic phanerogam populations of the *Orobanche* genus in Hungary. In: *Proceedings of the International Conference of Integrated Plant Protection,* Budapest, July 4-9, 1983, 4, 81–86.

Horvath, Z., Bujaki, G. Data on the host plant range and feeding habits of *Phytomyza orobanchia* Kalt. (Dipt., Agromyzidae) in Hungary [1990]. *Novenyvedelem (Hungary)*,1990, 26, 7, 296-301.

Horvath, Z., Osztrogonac, J. Regulating effect of goal 2E herbicide (Oxyfluorfen) on the populations of sunflower broomrape (*Orobanche cumana* Wallr.). *Novenyvedelem*, 1991, 27, 128-132.

Hsiao, A.I., Worsham, A.D., Moreland, D.E. Regulation of witchweed (*Striga asiatica*) conditioning and germination by dl-strigol. *Weed Sci.*, 1981, 29, 101-104.

Hunter, C., Poethig, R.S. Missing links: miRNA and plant development. *Curr. Opin. Genet. Dev.*, 2003, 13, 372–378.

Hussein, A.H.A., El-Deeb, M.A., Saleib, S.R. Effect of number and timing application of glyphosate on *Orobanche* infestation and faba bean yields in middle Egypt. *Arab Universities Journal of Agricultural Sciences*, 1998, 6, 437-444.

Ibrahim, T.S.I., Salim, A.A., El- Sherif, I., Abd El- Hamid, M.M., Shalaby, A.A. Demonstration plots for broomrape control in fababean. In: Americanos, P.G., Abu- Irmaileh, B.A., Saghir, A.R., (eds.), *Proceedings of the First Near East Conference on Improved Weed Management*, February, 2000. Khartoum, Sudan, 2000, pp. 186 -190.

ICARDA. Food legume improvement. *Progress Annual Report for 1986*. International Center for Ageicultural Research in Dry Areas, Aleppo, Syria, 1986, pp. 204-206.

ICARDA. *ICARDA Annual Report 1991*. International Centre for Agricultural Research in Dry Areas, Aleppo, Syria, 1992, 120pp.

ICARDA. Technology generations and dissemination for sustainable production of cereals and cool season legumes. International Center for Agricultural Research in the Dry Areas, Aleppo, Syria. 2006, pp. 256.

Imerovski, I., Dimitrijevic, A., Miladinovic, D., Dedić, B., Jocić, S., Tubić, N.B., Cvejic, S.D. Mapping of a new gene for resistance to broomrape races higher than F. *Euphytica*, 2016, 209, 281–289.

India, Central Tobacco Research Institute. Control of *Orobanche*, Leaflet, Central Tobacco Research Institute No.1 (revised), 1979, 8pp.

Irving, L.J., Cameron D.D. You are what you eat: interactions between root parasitic plants and their hosts. *Adv.Bot. Res.* 2009, 50, 88-138.

Ish-Shalom-Gordon, N., Jacobsohn, R., Cohen, Y. Inheritance of resistance to *Orobanche cumana* in sunflower. *Phytopathology*, 1993, 83, 1250-1252.

Ismail, A.M.A., Obeid, M. A study of assimilation and translocation in *Cuscuta hyaline* Heyne ex Roth. *Orobanche ramosa* L. and *Striga hermonthica* Benth. *Weed Res.,* 1976, 16, 87-92.

Jacobsohn, R., Ben-Ghedalia, D., Marton, K. Effect of animal's digestive system on the infectivity of *Orobanche* seeds. *Weed Res.*, 1987, 27, 87-90.

Jacobsohn, R., Dor, Z., Tanaami, Z., Ovdat, N. Crenate broomrape control in garden and field peas with foliarly applied imazethapyr. In: Wegmann, K., Musselman, L.J., Joel, D.M., (eds.), *Current Problems of Orobanche Researches. Proceedings of the Fourth International Orobanche Workshop,* Institute for Wheat and Sunflower, Albena, Bulgaria, 1998, pp. 331–336.

Jacobsohn, R., Greenberger, A., Katan, J., Levi, M., Alon, H. Control of Egyptian broomrape (*Orobanche aegyptiaca*) and other weeds by means of solar heating of soil by polyethylene mulching. *Weed Sci.*, 1980, 28, 312-316.

Jacobsohn, R., Kelman Y., Shaked, R., Klein, L. Control of broomrape in peas with ethylene dibromide and chlorpicrin. *Phytoparasitica*. 1982,10, 267. (Abstract).

Jacobsohn, R., Kelman, Y. Effectiveness of glyphosate in broomrape (*Orobanche* spp.) control in four crops. *Weed Sci.*, 1980, 28, 692-698.

Jacobsohn, R., Kelman, Y. Proper timing of glyphosate application for broomrape control in carrot and celery. *Phytoparasitica*, 1982, 10, 268. (Abstract).

Jafarzadeh, N., Pourmirza, A.A. A study on the biology of *Phytomyza orobanchia* Kalt. under laboratory and field conditions in Urmia (Iran). *Iranian Journal of Agricultural Sciences*, 1999, 30, 791-798.

Jain, R., Foy, C. L. Nutrient effects on parasitism and germination of Egyptian broomrape (*Orobanche egyptiaca*). *Weed Technol*., 1992, 6, 269- 275.

Jain, R., Foy, C. L. Parasitic flowering plants. In: Weber, H. Chr., Forstreuter, W., (eds.), *Proceedings of the 4th ISPFP*, Marburg, 1987 pp. 427-436.

Jinga, V., Iliescu, H., Ionita, A., Stanescu, V. Control of *Orobanche* on sunflower and tobacco crops in România (Jinga). *Joint Working Groups and MC Meeting of COST Action 849,* Broomrape Biology, Control and Management, September, 15-17, 2005, Reading University, UK. P. 40.

Joel, D.M., Aviv, D., Surov, T., Portnoy, V., Goldman-Guez, T., Gressel, J. Transformation of crops to herbicide-resistance and their use against parasitic weeds. In: *Plant Biotechnology and in Vitro Biology in the 21st century. Proceedings of the IXth International Congress of the International Association of Plant Tissue Culture and Biotechnology*, Jerusalem, June14-19,1998. Dordrecht, Netherlands; Kluwer Academic Publishers, 1999 pp. 499-502.

Joel, D.M., Chaudhuri, S.K., Plakhine, D., Ziadna, H., Steffens, J.C. Dehydro costus lactone is exuded from sunflower roots and stimulates germination of the root parasite *Orobanche cumana. Phytochemistry*, 2011, 72, 624–634. doi:10.1016/j.phytochemistry, 011.01.037

Joel, D.M., Kleifeld, Y., Losner-Goshen, D., Herzlinger, G., Gressel, J. Transgenic crops against parasites. *Nature*, 1995, 374, 220-221.

Joel, D.M., Losner-Goshen, D. The attachment organ of the parasitic angiosperms *Orobanche cumana* and *Orobanche aegyptiaca* and its development. *Can. J. Bot.*, 1994, 72, 564-574.

Joel, D.M. Key developmental processes in parasitic weeds as potential targets for novel control methods. In: Maillet, J. (ed.), *Comptes-rendus 6eme Symposium Mediterranean EWRS,* May 13-15,1998, Montpellier, France, 135-140.

Johnson, A.W., Roseberry, G., Parker, C. A novel approach to *Striga* and *Orobanche* control using synthetic germination stimulants. *Weed Res.*, 1976, 16, 223-227.

Jorrin, J., Ruck, E.D., Seghini, K., Perez, De-Luque A., Munoz-Garcia, J. Biochemical aspects of the parasitism of sunflower by *Orobanche*. In: Moreno, M.T., Cubero, J.I., Derner, D., Joel, D., Musselman, L.J., Parker, C. (eds.), *Advances in Parasitic Plant Research. Sixth International Parasitic Weed Symposium*, April 16-17-18, 1996, Cordoba, Spain, 1996 pp. 551-558.

Jurado-Exposito, M., Castejo′ n-Mun˜ oz, M., Garcı′a-Torre′ s, L., Uptake and translocation of imazethapyr in peas as affected by parasitism of *Orobanche crenata* and herbicide application methods. *Weed Res.*, 1999, 39, 129.

Jurado-Exposito, M., Castejon-Munoz, M., Gracia-torres, L. Broomrape (*Orobanche crenata* Forsk.) control with imazetapyr applied to pea (*Pisum sativum* L.) seeds. *Weed Technol.*, 1996, 10,774–781.

Jurado-Exposito, M., Gracia-Torres, L., Castejon-Munoz, M. Broad bean and lentil seed treatments with imidazolinones for the control of broomrape (*Orobanche crenata*). *J. Agric. Sci.*, 1997, 129, 307-314.

Kabulov, D.T. Broomrape of central Asia. Zarazikhovya srednei Azii. Tashkent, Uzbek SSR; Fan. 1978, 68 pp.

Kabulov, D.T., Kabulova, A.D. The anatomical structure of broomrape and the vegetative organs of parasitized plants. *Trudy Samarkandskogo Gosudarstvennogo Universiteta imeni A. Navoi*, 1977, 336, 92-108.

Kacan, K., Tursun, N. Effect of planting time and tomato varieties on broomrape (*Phelipanche aegyptiaca*) emergence and tomato yield in Western Turkey. *Res. Crops*, 2012, 13, 1070–1077.

Kamel, F. *Orobanche* situation in Algeria. *National Project Report*, 2005.

Kanai, A.; Natori, S. Cloning of gene cluster for sarcotoxin I, antibacterial proteins of *Sarcophaga peregrina. FEBS Lett.* 1989, 258,199–202.

Kapralov, S.I. *Phytomyza* against broomrape. *Zernovoe Chozjajstvo, Moskva,* 1974, 13(7), 43–44.

Kapralov, S.I. *Phytomyza* Einsatz im Gebiet Rostov. *Zascita Rastenij*, 1975, 20(6), 22–23.

Karnkowski, W. Notifiable weeds and parasitic-plants in the plant material imported into Poland during 1996-99. *Ochrona Roslin*, 1999, 43, 15-33.

Kasasian, L. Control of *Orobanche. Pest Article and News Summaries* (*PANS*), 1973, 19, 368-371.

Kasasian, L., Parker, C. The effect of numerous herbicides on the germination of *Orobanche aegyptiaca* and *Striga hermonthica. Pest Article and News Summaries* (*PANS*), 1971, 17, 471-481.

Kasasian, L. *Orobanche* spp. *Pest Article and News Summaries* (*PANS*), 1971, 17, 35-37.

Kasrawi, M.A., Abu-Irmaileh, B.E. Resistance to branched broomrape (*Orobanche ramosa*) in tomato germplasm. *HortScience*, 1989, 24, 822-824.

Kebede, M., Ayana, B. Economically important parasitic weeds and their management practices in crops. *Environ. Earth Sci.*, 2018, 8 (12), 104-115. 234665371. PDF.

Khalaf, K.A. Evaluation of the biological activity of flax as a trap crop against *Orobanche* parasitism of *Vicia faba. Trop.Agric.* 1992, 69, 35-38.

Khalaf, K.A., Ali, A.M., Youssef, K.A., Abd-Alaziz, S.A. Studies on the control of *Orobanche crenata.* I. Use of *Azotobacter* spp. and *Escherichia coli* transformants to break dormancy of *Orobanche crenata. FABISNewsletter*, 1997a, 40, 25-30.

Khalaf, K.A., Youssef, K.A., Ali, A.M., Abd-Alaziz, S.A. Studies on the control of *Orobanche crenata.* II. Effectiveness of *Azotobacter* spp. and *Escherichia coli* transformants in biological control of *Orobanche crenata* on *Vicia faba* under soil conditions. *FABIS Newsletter,* 1997b. 40, 31-35.

Khalil, D., Zemragh, M., Marliere, A.M. Broomrape control in Morocco: assessment and future prospects. In: *Compte Rendu de la 12e Conference du COLUMA*, 1993, Volume 1, pp. 195-202.

Khot, R. S., Bhat, B. N., Kadapa, S. N., Kambar, N. S. Effect of deep tillage in summer on *Orobanche* incidence and yield of bidi tobacco. *Tobacco Res.,* 1987, 13, 134-138.

Kim, H.I., Kisugi, T., Khetkam, P., Xie, X., Yoneyama, K., Uchida, K., Yokota, T., Nomura, T., McErlean, C.S.P., Yoneyama, K. Avenaol, a germination stimulant for root parasitic plants from *Avena strigosa. Phytochemistry*, 2014, 103, 85–88.

Kleifeld, Y. Progress in *Orobanche* control. In: Wegmann, K., Musselman, L.J., Joel, D.M. (eds.), *Current Problems of Orobanche Research. Proceedings of the 4th International Orobanche Workshop*. Albena, Bulgaria, 1998, pp. 327-330.

Kleifeld, Y. Role of pepper (*Capsicum annuum*) as a trap and catch crop for control of *Orobanche aegyptiaca* and *O. cernua. Weed Sci.*, 1966, 44, 948-951.

Kleifeld, Y., Goldwasser Y., Herzlinger, G., Golan, S., Baraghutti, A. Broomrape control with metham–sodium. In: Ransom, J.K., Musselman, L.J., Worsham, A.D., Parker, C., (eds.), *Proceedings of the 5th International Symposium of parasitic Weeds*, 24-30 June 1991, Nairobi, Kenya,

Kleifeld, Y., Goldwasser, Y., Herzlinger, G., Joel, D.M., Golan, S., Kahana, D. The effects of flax *Linum usitatissimum* (L.) and other crops as trap and catch crops for control of Egyptian broomrape (*Orobanche aegyptiaca* Pers.). *Weed Res.*, 1994, 34, 37-44.

Kleifeld, Y., Goldwasser, Y., Plakhine, D., Eizenberg, H., Herzlinger, G., Golan, S. Selective control of *Orobanche* spp. in various crops with sulfonylurea and imidazolinones herbicides. In: *Proceedings of Regional Workshop on Joint Action to Control Orobanche in the WANA-region: Experiences from Morocco*. 1999, Rabat, Morocco, GTZ GmbH., pp 26.

Kleifeld, Y., Herzlinger, G. Chemical control of broomrape in sunflowers. *Hassadeh*, 1984, 64, 2011.

Kleifeld, Y., Regev, Y., Herzlinger, G., Bargutti, A. Experiments in chemical control of broomrape. 1982 (Weed Abstract, 1984, 33, 4, Abstract no. 914).

Klein, O. *Untersuchungen zur Populationsdynamik und zur Verwendung von* Phytomyza orobanchia in der *Biologischen Bekämpfung* von Orobanche spp. *in Marokko*. M.Sc. Thesis, Department of Plant Production and Agroecology in the Tropics and Subtropics, University of Hohenheim, Germany. 1995, 99 pp.

Klein, O., Kroschel, J. Biological control of *Orobanche* spp. with *Phytomyza orobanchia:* a review. *BioControl*, 2002, 47, 245-277.

Klein, O., Kroschel, J., Sauerborn, J. Potential of *Phytomyza orobanchia* for the biological control of *Orobanche* spp. and its possible application. In: Cubero, J.I., Moreno, M.T., Rubiales, D., Sillero, J.C., (eds.), *Resistance to Broomrape, the State of Art*, Junta Andalucia, Sevilla, Spain, 1999, pp. 179-185.

Klyueva, M.P., Bonduryanskaya, B.J. Effect of *Phytomyza orobanchia* Kalt. on the seed productivity of broomrape. In: Shapa, V.A., (ed.), *Biological Plant Protection,* 1986, 45, 39–44.

Klyueva, M.P., Pamukchi G.V., Ursake G.P., Bonduryanskaya, B.J. *Rasprostranenie podsolnuchovoj i vetvistoj sarasich v moldavii i sarajennost' ich fitomisoj. Zascita Rastenij, Kishinev*. 1972, pp. 54–59.

Klyueva, M.P., Pamukchi, G.V. Broomrape midge, the natural enemy of broomrape in Moldova.) *Istvestiya Akademii Nauk Moldavskoi SSR, Seriya Biologicheskikh I Khimicheskikh Nauk (USSR)* 1978, 4, 21–25.

Klyueva, M.P., Pamukchi, G.V. Trofitscheskie svjasi *Phytomyza orobanchia* Kalt. (Diptera: Agromyzidae). *Biologicheskikh Metodov Zascita Rastenij*, 1979. pp. 23–25.

Klyueva, M.P., Ursake, G.P. The number of *Phytomyza* in flowers and stems of *Orobanche*. Entomofagi, *Fitofagi i Mikroorganizmi v Zascita Rastenij, Kishinev*. 1974, pp. 34–39.

Kott, S.A., Biological control of broomrape. Sornye rastenija i bor'ba s nimi. In: *Weeds and Their Control*. Kolos, Moskva, USSR. 1969, pp. 169–171.

Kovačević, J. Korovi u poljoprivredi. Nakladni zavod Znanje, Zagreb. 1976.

Krishnamurthy, G.V.G. *Orobanche* problem and management in India. In: Pieterse, A.H.; Verkleij, J.A.C., ter Borg, S.J., (eds.), *Biology and Management of Orobanche. Proceedings of the Third International Workshop on* Orobanche *and Related* Striga *Research*, November 8-12,1993. Amsterdam, Netherlands; *Royal Tropical Institute* 1994, pp. 591-598.

Krishnamurthy, G.V.G., Chandwani G.H. Effect of various crops on the germination of *Orobanche* seeds. *Pest Article and News Summaries* (*PANS*), 1975, 21, 64-66.

Krishnamurthy, G.V.G., Chari, M.S. Further studies on post-emergence control of broomrape on tobacco with plant oils. In: Ransom, J.K., Musselman, L.J., Worsham, A.D., Parker, C., (eds.), *Proceedings of the Fifth International Symposium on Parasitic Weeds*, Nairobi, Kenya: CIMMYT, 1991, pp. 108-110.

Krishnamurthy, G.V.G., Lal, R., Nagarajan, K. Further studies on the effect of various crops on the germination of *Orobanche* seed. *Pest Article and News Summaries* (*PANS*), 1977, 23, 206-208.

Krishnamurthy, G.V.G., Nagarajan, K. Post-emergence control of *Orobanche* on tobacco with oils. *Trop. Pest Manag.,* 1991b, 37, 149-151.

Krishnamurthy, G.V.G., Nagarajan, K., Chandwani, G.H. Studies on the control of *Orobanche* on tobacco. *Tobacco Res*earch, 1976, 2, 58-62.

Krishnamurthy, G.V.G., Nagarajan, K., Lai, R. Studies on longevity and depth of emergence of broomrape in tobacco fields. *Indian Journal of Weed Science*, 1987, 17, 38-41.

Krishnamurthy, G.V.G., Rao, M. Control of *Orobanche* through rotation. *Indian Farming*, 1976, 25, 23.

Krishnamurthy, G.V.G. Two new methods to control *Orobanche* on tobacco. *Indian Farming*, 1992, 42, 7-8.

Kroschel, J., Jost, A., Sauerborn, J. Insects for *Striga* control: Possibilities and constraints, in: Kroschel J., Mercer-Quarshie H., Sauerborn J. (eds.*), Advances in Parasitic Weed Control at On-Farm Level. Vol. I. Joint Action to Control Striga in Africa*. Margraf Verlag, Weikersheim, Germany, 1999, pp. 117-132. *(17) (PDF) Innovations in parasitic weeds management in legume crops. A review*. Available at: https://www.researchgate.net/publication/225133474_Innovations_in_parasitic_weeds_management_in_legume_crops_A_review#fullTextFileContent [accessed Jul 04 2021].

Kroschel, J. Integrated pest management in potato production in the Republic of Yemen with special reference to the integrated biological control of the potato tuber moth (*Phthorimaea operculella* Zeller). *Tropical Agriculture* , 1995, 8. Weikersheim, Germany: Margraf Verlag.

Kroschel, J., Mueller-Stoever, D., Elzein, A., Sauerborn, J. The development of mycoherbicides for the management of parasitic weeds of the genus *Striga* and *Orobanche* - a review and recent results. In: Spencer, N.R., (ed.)*, Proceedimgs of the Xth Int. Symposium on Biological Control of Weeds,* July4-14,1999, Bozeman, Montana, USA. 2000, pp. 139.

Kruh, L.I., Lahav, T., Abu-Nassar, J., Achadri, G., Salami, R., Freilich, S., Aly, R. Host-parasite bacteria triangle: the microbiome of the parasitic weed *Phelipanche aegyptiaca* and tomato-*Solanum lycopersicum* (Mill.) as a host. *Front. Plant Sci*., 2017, (cited 2017 Dec 14); (9 p.). DOI:10.3389/fpls.2017.00269

Kukula, S.T., Masri, H. Integrated cultural practices and chemical control of *Orobanche crenata* in faba bean. In: Parker, C., Musselman, L.J., Polhill, R. M., Wilson, A.K., (eds.), *Proceedings, 3rd International Symposium on Parasitic Weeds*, Aleppo, Syria, ICARDA, Aleppo, 1984, pp. 256-261.

Labrada, R., Perez, R. Non-chemical control methods for *Orobanche ramosa* (in Spanish). *Agrotecnica de Cuba*, 1988, 20, 35-40.

Labrousse, P., Zehhar, N., Arnaud, M.C., Boulet, C., Fer, A., Thalouan, P. Crops, wild species and weeds present similar defense reactions to various root parasitic angiosperms. In: Fer, A., Thalouarn, P., Joel, D.M., Musselman, L.J., Parker, C., Verkleij, J.A.C., (eds.), *Proceedings of the 7th International Parasitic Weed Symposium*, Nantes, France, 2001, 237.

Lande, T., Hershenhorn, J., Achdari, G., Eizenberg, H. New advances in chemical control of *Orobanche aegyptiaca* in tomato (Lande). *Joint Working Groups and MC meeting of COST Action 849, Broomrape Biology, Control and Management,* September 15-17, 2005, Reading University, UK. 37.

Lekic, M.B. Investigations of the dipteran *Phytomyza orobanchia* Kalt. as a controller of parasitic phanerogams of the genus *Orobanche*. *Savremena Poljoprivreda, Novi Sad,* 1974, 22(1–2), 93–99.

Lekic, M.B. Phytophagous insects observed on parasitic phanerogams of the genera *Orobanche* and *Cuscuta* in 1968. In: Simmonds, F.J., (ed.), *Proceedings, 1st International Symposium on Biological Control of Weeds,* Delemont, 1969. *Commonwealth Institute of Biological Control, Miscellaneous Publication*, 1970a, 1, 21-24.

Lekic, M.B. The role of the dipteran *Phytomyza orobanchia* Kalt. (Agromyzidae) in reducing parasitic phanerogam populations of the *Orobanche* genus in Vojvodina. *Contemporary Agriculture*, 1970b, 18(7–8), 59–68.

Linke, K.H., Abd El-Moneim, A.M., Saxena, M.C. Variation in resistance of some forage legumes species to *Orobanche crenata* Forsk. *Field Crops Res.*, 1993, 32, 277-285.

Linke, K.H., Saxena, M.C. Towards an integrated control of *Orobanche* spp. in some legume crops. In: Wegmann, K., Musselman, L.J., (eds.), *Proceedings of International Workshop in* Orobanche *Research,* Tubingen, Germany: *Eberhard-Karls-Universität*, 1991, pp. 248-256.

Linke, K.H., Scheibel, C., Saxena, M.C., Sauerborn, J. Fungi occurring on *Orobanche* spp. and their preliminary evaluation for *Orobanche* control. *Trop. Pest Manag.*, 1992, 38, 127-130.

Linke, K.H., Schnell, H., Saxena, M.C. Factors affecting the seed bank of *Orobanche crenata* in fields under lentil based cropping systems in northern Syria. In: Ransom J.K.; Musselman L.J.; Worsham A.D., Parker, C., (eds.), *Proceedings, 5th International Symposium on Parasitic Weeds.* CIMMYT, Nairobi, 1991, pp. 321-327.

Linke, K.H., Vorländer, C., Saxena, M.C. Occurrence and impact of *Phytomyza orobanchia* (Diptera: Agromyzidae) on *Orobanche crenata* (Orobanchaceae) in Syria. *Entomophaga,* 1990, 35(4), 633–639.

Lolas, P. C. Control of broomrape (*Orobanche ramosa*) in tobacco (*Nicotiana tabacum*). *Weed Sci.*, 1986, 34,427-430.

Lolas, P. Herbicides for control of broomrape (*Orobanche ramosa* L.) in tobacco (*Nicotiana tabacum* L.). *Weed Res.*, 1994, 34, 205-209.

Lolas, P. Sub-group collaborative study on broomrape. 1996-1997 Report. *Bulletin d'Information CORESTA*, 1997, 3, 79-85.

Lolas, P.C. Methods and strategies for control of broomrape in tobacco. In: *Proccedings of 1998 Tobacco Symposium, Indian Tobacco-Problems and Prospects*, Rajahmundry, India, 20-23 January 1998. Indian Society of Tobacco Science, 1998, pp. 33-42.

Lopez-Granados, F., Diaz-Sanchez, J., Garcia-Torres, L. A bioeconomic model for broomrape control (*Orobanche crenata*) in faba beans (*Vicia faba*) under different management strategies. In: *Proceedings of the 1997 Congress of the Spanish Weed Science Society*, Valencia, Spain, November 24-26, 1997, Madrid, Spain.

Lopez-Granados, F., Garcia-Torres, L. Effect of environmental factors on dormancy and germination of crenate broomrape (*Orobanche crenata*). *Weed Sci.*, 1996, 44, 284-289.

Lopez-Raez, J.A., Charnikhova, T., Fernandez, I., Bouwmeester, H., Pozo, M.J. Arbuscular mycorrhizal symbiosis decreases strigolactone production in tomato. *J. Plant Physiol.* 2011,168, 294–297

Louarn, J., Boniface, M-C, Pouilly, N., Velasco, L., Pérez-Vich, B., Vincourt, P., Muños, S. Sunflower resistance to broomrape (*Orobanche cumana*) is controlled by specific QTLs for different parasitism stages. *Front. Plant Sci.*, 2016, (14 p.). DOI: 10.3389/ fpls. 2016.00590

Louarn, J., Carbonne, F., Delavault, P., Be´card, G., Rochange, S. Reduced germination of *Orobanche cumana* seeds in the presence of arbuscular mycorrhizal fungi or their exudates. *PLoS ONE*, 2012, 7(11), e49273. https://doi.org/10.1371/journal.pone.0049273.

Lovett, J.V., Fraser, S.A., Duffield, A.M. Allelopathic activity of cultivated sunflowers. In: *Proceedings 10th International Sunflower Conference*, Australian Sunflower Association, International Sunflower Association, Surfers' Paradisi, 1982, pp.198-201.

Lu, Y., Gagne, G., Grezes-Besset, B., Blanchard, P. Integration of a molecular linkage group containing the broomrape resistance gene Or5 into an RFLP map in sunflower. *Genome,* 1999, 42, 453- 456.

Luque, A.P.De, Galindo, J.C.G., Macias, F.A., Jorrin, J. Sunflower sesquiterpene lactone models induce *Orobanche cumana* seed germination. *Phytochemistry*, 2000, 53, 45-50.

Lyra, S., Economou, G., Anastasiadou, A., Meletiou, A., Kotoula- Syka, E. Biocontrol agents for *Orobanche* - A seaweed product: a new potential germination stimulant for *Orobanche ramosa*

(Economou). 2005. *Joint Working Groups and MC 0eeting of COST Action 849, Broomrape Biology, Control and Management,* 15-17 September 2005, Reading University, UK.

M¨uller-St¨over, D., Kroschel, J., Thomas, H. Sauerborn, J. Chlamydospores of *Fusarium oxysporum* Schlecht f. sp. *orthoceras* (Appel and Wollenw.) Bilai as inoculum for wheat-flour–kaolin granules to be used for the biological control of *Orobanche cumana* Wallr. *Eur. J. Plant Pathol.,* 2002,108, 221–228.

Ma, Y. Q., Zhang, W., Dong, S.Q., Ren, X. X., An, Y., Lang, M. Induction of seed germination in *Orobanche* spp. by extracts of traditional Chinese medicinal herbs. *Sci. China Life Sci.*, 2012, 55, 250–260, doi: 10.1007/s11427-012-4302-2.

Mabrouk, Y., Belhadj, O. Effect of the inoculation of chickpea by rhizobia on growth promotion and protection against *Orobanche crenata. Global Journal of Biology, Agriculture and Health Science's,* 2014, 3(3), 55–59.

Mabrouk, Y., Mejri, S., Hemissi, I., Belhadj, O. Biochemical analysis of induced resistance in chickpea against broomrape (*Orobanche foetida*) by Rhizobia inoculation. *Phytopathol. Mediterr.,* 2016, 55, 1, 54–61. DOI: 10.14601/Phytopathol_Mediterr-16338.

Mabrouk, Y., Zourgui, L., Sifi, B., Belhadj, O. The potential of *Rhizobium* strains for biological control of *Orobanche crenata. Biologia, Bratislava* (*Section Botany*), 2007, 62(2), 139—143, DOI: 10.2478/s11756-007-0021-8.

Mamonov, I..F., Poplaukhin, V.P., Nazarko, T. The results of sunflower selection for resistance to broomrape. *Selektsiya I Semenovodstvo, USSR*, 1978, 1, 23-25.

Manjunath, T.M., Nagarkatti, S. Natural enemies of *Orobanche* in India and possibilities of its biological control. *Technical Bulletin of the Commonwealth Institute of Biological Control,* 1977, 18, 75-83.

Manschadi, A. M., Sauerborn, J., Stutzel, H. Quantitative aspects of *Orobanche crenata* infestation in faba beans as affected by abiotic factors and parasite soil seedbank. *Weed Res.*, 2001, 41, 311-321.

Mariam, *E.G.*, Suwanketnikom, R. Effect of nitrogen fertilizers on branched broomrape (*Orobanche ramosa* L.) in tomato (*Lycopersicon esculentum* Mill.). *Kasetsart J.* (*Nat. Sci.*), 2004, 38, 311–319. Available at: http://ext.agn.uiuc.edu/wssa/subpages/weed/WT72.htm

Martelli, G.M. Nota preliminare sui parassiti animali dell *Orobanche* della Fava *Orobanche speciosa* D.C. *Riv. Patol.Veg.,* 1933a, 23, 233–240.

Martelli, G.M. Primo contributo alla conoscenza di alcuni parassiti dell'*Orobanche* della Fava (*Orobanche speciosa* D.C.). *Bolletino del Laboratorio de Zoologia Generale e Agraria in Portic'i,* 1933b, 28, 11–36.

Matthews, D.E., Joel, D.M., Steffens, J.C. Species-specific response of *Orobanche* to host-derived and synthetic germination stimulants. In: Ransom, J.K., Musselman, L.J., Worsham, A.D., Parker, C., (eds.), *Proceedings of the 5th International Symposium of Parasitic Weeds*, June 24-30, 1991, Nairobi, Kenya.

Matthews, J.M. Herbicide and cropping trails releventto eradication of branched broomrape (*Orobanche ramosa*) in South Australia. In: Spafford Jacob, H., Dodd, J., Moore, J.H. (eds.), *Proceedings 13th Australian Weed Conference* Perth. 2002, 274-275.

Mauro, R.P., Monaco, A. L., Lombardo, S., Restuccia, A., Mauromicale, G. Eradication of *Orobanche/Phelipanche* spp. seed bank by soil solarization and organic supplementation. *Sci. Hortic.*, 2015, 193, 62–68.

Mauromicale, G., Lo monako, A., Longo, A. Effect of branched broomrape (*Orobanche ramosa*) infection on the growth and photosynthesis of tomato. *Weed Sci.*, 2008, 56(4), 574-581.

Mazaheri, A., Moazami, N., Vaziri, M., Moayed-Zadeh, N. Investigation of *Fusarium oxysporum* a possible biological control of broomrape (*Orobanche* spp.). In: Ransom, J.K., Musselman,

L.J., Worsham, A.D., Parker, C., (eds.), *Proceedings of the 5th International Symposium of Parasitic Weeds*, June 24-30, 1991, Nairobi, Kenya.

Mesa- Garcia, J., Garcia- Torres, L. *Orobanche crenata* (Forsk) control in *Vicia faba* (L.) with glyphosate as affected by herbicide rates and parasite growth stages. *Weed Res.*, 1985, 25, 129-134.

Mesa-Garcia, J., Garcia-Torres, L. Broomrape (*Orobanche crenata* Forsk.) control in bean (*Vicia faba* L.) with glyphosate affected by infection intensity. In: *Proceedings of British Crop Protection Conference-Weeds*, Brighton, 1982, pp. 765-770.

Metha, H.M., Trivedi, G.C., Patel, H.R. Study to find out effective method of weed control in bidi tobacco. In: *Abstracts of papers, Annual Conference of Indian Society of Weed Science*, 1985.

Meti, S.S., Hosmani, M.M. Broomrape control in bidi tobacco by soil solarization. *Tobacco Research*, 1994, 20, 60-70.

Miccolis, V., Bianco, V.V. The influence of glyphosate, propyzamide and some cultural techniques on the infestation of *Orobanche crenata* Forsk. in broad beans for fresh consumption. *Difesa delle Piante*, 1984, 7, 161-175.

Mihajlovic, L. Results of investigation on *Orobanche* spp. entomofauna in Yogoslavia and the possibility of using insects for biological control. In: terBorg, S.J., (ed.), *Proceedings, Workshop on Biology and Control of* Orobanche, Wageningen, LH/PVO, Wageningen, 1986 pp. 118-126.

Miladinovic, D., Dedic, B., Quiroz, F. *Orobanche cumana* Wallr. Resistance of commercial sunflower cultivars grown in Argentina. *J. Basic Appl. Genet.*, 2012, 23, 37–41.

Miller, A.E., Douce, G.K., Murphy, T.R., Waston, J.T. Cooperative extension. Department of Entomology, *Cooperative Extension Service College of Agricultural and Environmental Sciences,* the University of Georgia, April 1997 GACAPS 0497.

Mohamed-Ahmed, A.G., Drennan, D.S.H. Factors affecting establishment of *Orobanche* spp. in legumes. In *Biology and Management of Orobanche.* Pieterse, A.H., Verkleij, J.A.C., ter Borg, S.J., (eds.), *Proceedings of the third International Workshop on Orobanche and Related Striga Research*, November 8-12, 1993, Amsterdam, Netherlands. 1994.

Moiseeva, N., Mamraliev, J. Zarazichu mozno pobedit (Broomrape (*Orobanche* sp.) can be eaten by *Phytomyza* sp.). *Sel'skoe Khozyaistvo Kirgizii*, 1969, 7, 24.

Mu¨ller-Sto¨ Ver, D, Kroschel, J. The potential of *Ulocladium botrytis* for biological control of *Orobanche* spp. *Biol. Control*, 2005, 33, 301–306.

Mu¨ller-Sto¨ Ver, D. Possibilities of biological control of *Orobanche* and *O. cumana* with *Ulocladium botrytis* and *Fusarium oxysporum* f. sp. *orthoceras*. *Agroecology* 3. Apia Verlag, Laubach, Germany, 2001, 174 pp.

Mu¨ller-Sto¨ Ver, D., Buschmann, H., Sauerborn, J. Increasing control reliability of *Orobanche cumana* through integration of a biocontrol agent with a resistance-inducing chemical. *Eur. J. Plant Pathol.*, 2005, 111, 193–202.

Mu¨ller-Sto¨ Ver, D., Kohlschmid, E., Sauerborn, J. A novel strain of *Fusarium oxysporum* from Germany and its potential for biocontrol of *Orobanche ramosa*. *Weed Res.*, 2009, 49, 175–182. DOI: 10.1111/j.1365-3180.2008.00686.x.

Mu¨ller-Sto¨ Ver, D., Thomas, H., Sauerborn, J., Kroschel, J. Two granular formulations of *Fusarium oxysporum* f. sp. *orthoceras* to mitigate sunflower broomrape *Orobanche cumana*. *BioControl,* 2004, 49, 595–602.

Mukumov, Kh. M., Faizieva, S. Z. Results of container trials investigating the broomrape resistance of vegetable and melon crops. *Trudy Samarkandskogo Gosudarstvennogo Universiteta imeni A. Navoi*, 1977, 336, 73-78.

Mukumov, Kh. The degree of germination of broomrape seed in relation to the growth stage of host plants. *Nauchnye Trudy Biologicheskogo Fakulteta, Samarkandskii Gosudarstennye University, lmeni A. Navoi Botanika,* 1974, 207, 134-140.

Munoz, M.C., Munoz, F.R., Torres, L.G. Effect of planting date on broomrape *Orobanche cernua* Loefl. infection in sunflower (*Helianhus annuus* L.). *Weed Res*., 1993, 33, 171-176.

Musselman L.J. The biology of *Striga, Orobanche*, and other root-parasitic weeds. *Annu. Rev. Phytopathol.*, 1980, 18, 463-489.

Musselman, L.J. 1998. *Checklist of plants of the Hashemite Kingdom of Jordan*. Old Dominion University. Available at: http://www.odu.edu/~lmusselm/plant/jordan/index.php

Musselman, L.J. Alien broomrape (*Orobanche* species) of potential danger to American crops. In: *Proceedings Southern Weed Science Society*: *Weed Science in Harmony with the Environment, 46th Annual Meeting*, Charlotte, North carolina, USA, January18-20, 1993. Champaign, Illinois, USA; *Southern Weed Science Society, Department of Biological Sciences, Old Dominion University, Norfolk, VA 23529, USA*. 1993 pp. 244-247.

Nadal, S., Moreno, M.T., Roma´n, B. Control of *Orobanche crenata* in *Vicia narbonensis* by glyphosate. *Crop Prot*., 2008, 27, 873–876.

Nandula, V.K., Foy, C.L., Orcutt, D.M. Glyphosate for *Orobanche aegyptiaca* control in *Vicia sativa* and *Brassica napus*. *Weed Sci*., 1999, 47, 486-491.

Nandula, V.K., Foy, C.L., Westwood, J.H. Environmental influences on germination of *Orobanche*. In: Moreno, M.T., Cubero, J.I., Berner, D.K., Joel, D., Musselman, L., Parker, C., (eds.).*Advances in Parasitic Plant Research, Proceedings, Sixth Parasitic Weed Symposium*, Cordoba, Spain.16-18 April, 1996, pp., 409-416.

Nassib, A.M., Hassanein, E.E., Hussein, A.H.A. Broomrape control in faba (*Vicia faba* L.) with reduced rates of glyphosate mixed with three adjuvants. *Ann. Agric. Sci. Moshtohor*, 1990, 28, 17-25.

Nassib, A.M., Hussein, A.H., El Rayes, F.M. Effect of variety, chemical control, sowing date and tillage on *Orobanche* spp. infestation and faba bean yield. *FABIS Newsletter*, 1984, 10, 11-15.

Nativ, N., Hacham. Y., Hershenhorn, J., Dor, E., Amir, R. Metabolic investigation of *Phelipanche aegyptiaca* reveals significant changes during developmental stages and in its different organs. *Front Plant Sci*., 2017, 13 p. DOI:10.3389/fpls.2017.00491.

Nefkens, G.H., Thuring, J.W.J.F., Beenakkers, M.F.M., Zwanenburg, B. Synthesis of a phthaloylglycine-derived strigol analogue and its germination stimulatory activity toward sees of the parasitic weeds *Striga hermonthica* and *Orobanche crenata*. *J. Agric. Food Chem.*, 1997, 45(6), 2273-2277.

Nefzi, F., Trabelsi, I., Amri, M., Triki, E., Kharrat, M., Zouhaier, Abbes, Z. Response of some chickpea (*Cicer arietinum* L.) genotypes to *Orobanche foetida* Poir. parasitism. *Chil. J. Agric. Res*., 2016, 76,170–178.

Nemat Alla, M.M., Shabana, 1.Y.M., Serag, M.M., Hassan, N. M. and El-Hawary, M.M. Granular formulation of *Fusarium oxysporum* for biological control of faba bean and tomato *Orobanche*. *Pest Manag. Sci.,* 2008, 64, 1237–1249.

Nemli, Y., Giray, H. Investigations on natural control of broomrape (*Orobanche* sp.) by *Phytomyza orobanchia* Kalt. (Diptera, Agromyzidae) in Izmir (Turkey). *Journal of Turkish Phytopathology*, 1983, 12, 39–44.

Nevena, B., Senka, D. *Orobanche ramosa* as a new host for the epiphytic stage of *Pseudomonas syringae* pv. *tomato,* Plant Protection Institute 2230 Kostinbrod, Bulgaria, 1997 pp. 67-68.

Nickrent, D.L. Parasitic plants. In: *McGraw-Hill Yearbook of Science & Technology*, 2008, 251-253.

Nickrent, D.L., Musselman, L.J. Introduction to parasitic flowering plants. *The Plant Health Instr.*, 2004. DOI: 10.1094/PHI-I-2004-0330-01.

Okazova, A. G., Vasyura, V. P., Okazov, P. N. Ploughing deep decreases broomrape infestation of tobacco. *Zasch. Rast.* (*Mosc.*), 1980, 9, 21.

Oliveira Velloso, J. *El Genero Orobanche en los Cultivos en el sur de Espaha, Taxonomie, Corologia Ecologia.* Tesis doctoral, Univesidada de Cordoba. Diciembre de 1990.

P′erez-de-Luque, A., Rubiales, D. Nanotechnology for parasitic plant control. *Pest Manag Sci.*, 2009, 65, 540–545.

Palled, Y.B., Path Y.S.; Panchal Y.C., Hosmani, M.M. Studies on chemical weed control in irrigated bidi tobacco (*Nicotiana tabacum* L.) with special reference to *Orobanche* (*Orobanche cernua* Loefl.). *Indian Journal of Weed science*, 1985, 17, 6-11.

Papyan, A.G., Arutyunyan, Kh. M. Harmful effect of broomrape on tobacco plants. *Izvestiya Sel'skokhozyaistvennoi Nauki Armyanskoi SSR*, 1981, 5, 33-37.

Parker, C. Observations on the current status of *Orobanche* and *Striga* problems worldwide. *Pest Manag. Sci.,* 2009, 65, 453–459. Available at: www.interscience.wiley.com. DOI 10.1002/ps.1713.

Parker, C. Parasitic weeds: a world challenge. *Weed Sci.*, 2012, 60, 269–276. Parker, C. Protection of crops against parasitic weeds. *Crop Prot.*, 1991, 10, 6-22.

Parker, C., Riches, C.R. *Parasitic Weeds of the World, Biology and Control.* CAB International, Wallingford, UK. 1993, 350 pp.

Pavan, S., Schiavulli, A., Marcotrigiano, A.R., Bardaro, N., Bracuto, V., Francesca Ricciardi, F., Charnikhova, T., Lotti, C., Bouwmeester, H., Ricciardi, L. Characterization of low-strigolactone germplasm in pea (*Pisum sativum* L.) resistant to crenate broomrape (*Orobanche crenata*Forsk.). *Mol. Plant Microbe Interact.*, 2016, 29,743–749.

Perez-de-Luque, A., Rubiales, D., Galindo, G.C., Macias, F.A., Jorrin, J. Allelopathy and allelochemicals within the plant-parasitic weed interaction. Studies with the sunflower-*Orobanche cumana* system. In: Fer, A., Thalouarn, P., Joel, D.M., Musselman, L.J., Parker, C., Verkleij, J.A.C., (eds.), *Proceedings of the 7th International Parasitic Weed Symposium*, Nantes, France, 2001, 196-199.

Pesch, C., Pieterse, A.H. Inhibition of germination in *Striga* by means of urea. *Experientia*, 1982, 38, 559-560.

Petzoldt, K., Nemli, Y., Sneyd, J. Integrated control of *Orobanche cumana* in sunflower. In: *Biology and Management of Orobanche.* Pieterse, A.H.; Verkleij, J.A.C., ter Borg, S.J., (eds.), *Proceedings of the Third International Workshop on* Orobanche *and Related* Striga *Research*, November 8-12, 1993, Amsterdam, Netherlands.

Pieterse, A. H. Research on broomrape and *Striga* at the Royal Tropical Institute and Cooperating Institute. In: Kim, S.K., (ed.), *Combating Striga in Africa, Proceedings of the International Workshop*, Ibadan, Nigeria, International Institute of tropical Agriculture, 1991b, August 22-24, 1988 Ibadan, Nigeria, pp. 69-72.

Pieterse, A. H. The effect of nitrogen fertilizers on the germination of seeds of *Striga hermonthica* and *Orobanche crenata.* In: Wegmann, K., Musselmann, L.J. (eds.), *Progress in Orobanche Research*, 1991a. Eberhard-karls-Universitat, Tubengen, FRG., pp. 115-124.

Pieterse, A.H. Germination of *Orobanche crenata* Forsk. seeds *in vitro. Weed Res.*, 1981, 21, 279-287.

Pieterse, A.H., Verkleij, J.A.C. Genetic variability in *Orobanche* (broomrape) and *Striga* (witchweed) and its implications for host resistance breeding. In: Rozema, J., Verkleij, J.A.C. (eds.), *Ecological Responses to Environmental Stresses*, Dordrecht, Netherlands; Kluwer Academic Publishers, 1991, pp. 209-302.

Pirani, V., Capriotti, G., Del Gatto, A., Tiberi, M. Sowing date, anti-cryptogam treatments, *Orobanche* damage and losses at harvesting in dry pea crops. *Informatore Agrario*, 1993, 49, 30-37.

Popova, A.A. The effect of fertilizers on the infection of *Nicotiana rustica* with wildfire and *Orobanche ramosa* L. Vses. Inst. Tabachn. I Makhov. *Promyshl. Krasnodar*, 1942, 144, 158-175.

Pouvreau, J.B., Gaudin, Z., Auger, B., Lechat, M.M., Gauthier, M., Delavault, P., Simier, P. A high throughput seed germination assay for root parasitic plants. *Plant Methods*, 2013, 9, 32. http://www.plantmethods.com/content/9/1/32.

Prider, J., Williams, A. Using dazomet to reduce broomrape seed banks in soils with low moisture content. *Crop Prot*. 2014, 59, 43–50.

Punia, S.S. Biology and control measures of *Orobanche. Indian Journal of Weed Science,* 2014, 46(1), 36–51.

Punia, S.S., Yadav, A., Singh, S., Sheoran, P., Yadav, D.B., Yadav, B. Broomrape: A threat to mustard cultivation in Haryana and its control measures, pp. 105. In: *Proceedings of 1st Brassica Conference "Production Barriers and Technological options in Oilseeds* Brassica" March 2-3, 2012, CCS HAU, Hisar, India.

Punia, S.S., Yadav, A., Yadav, D.B., Singh, S. Management of *Orobanche aegyptiaca* in Indian mustard, In: *Proceedings of Biennial Conference of ISWS "Recent Advances in Weed Science-*2010" February 25-26, 2010, IGKVV, Raipur (Chhatisgarh). pp. 174.

Pusch, J., Gu¨nther, K.F. Orobanchaceae (Sommerwurzgewa¨chse).In: Hegi, G., (ed.), *Illustrierte Flora von Mitteleuropa Bd. 6/1A, Lieferung 1.* Weissdorn, Jena, 2009 pp. 1–99.

Putnam, A.R., Duke, W.B. Allelopathy in agro-ecosystem. *Ann. Rev. Phytopathol.*, 1978, 16, 431-457.

Puzzilli, M. Tobacco broomrapes and their control and some useful references to other parasite and host species. *Review of Agriculture Subtropicale e Tropicale,* 1983, 78, 209-248.

Qasem J.R. Plants as sources of natural herbicides against branched broomrape (*Orobanche ramosa* L.). In: Reigosa, M. J., Pedrol, N. (eds.), *Allelopathy: From Molecule to Ecosystem*, Science Publisher, Inc. Enfield, USA, 2002, 153-182.

Qasem, J. R. Parasitic weeds and allelopathy, from the hypothesis to the proof. In: Reigosa, M. J., Pedrol, N., Gonzalez, L. (eds.), *Allelopathy, a Physiological Process with Ecological Implications*, 2006b, pp. 565- 637). Springer, the Netherlands.

Qasem, J. R. Chemical control of branched broomrape (*Orobanche ramosa*) in glasshouse grown tomato. *Crop Prot.*, 1998, 17(8), 625-630.

Qasem, J. R. Parasitic weeds of the Orobanchaceae family and their natural hosts in Jordan. *Weed Biol. Manag.*, 2009, 9 (2), 112-122.

Qasem, J. R. *The Coloured Atlas of Medicinal and Aromatic Plants of Jordan and Their Uses. Volume Three*. Cambridge Scholars Publishing. NewCatle, UK. 2020b, 491 PP. ISBN 978-1-5275-4561X

Qasem, J.R. Branched broomrape (*Orobanche ramosa* L.) control in tomato (*Lycopersicon esculentum* Mill.) by trap crops and other plant species in rotation. *Crop Prot.*, 2019, 120, 75–83.

Qasem, J.R. Control of branched broomrape (*Orobanche ramosa* L.) in tomato (*Lycopersicon esculentum* Mill.) by olive cake and olive mill waste water. *Crop Prot.*, 2020a, 129 105021. DOI: https://doi.org/10.1016/j.cropro.2019.105021.

Qasem, J.R. Herbicide resistant weeds: The technology and weed management. In: Price, A.J., Kelton, J.K., (eds.), *Herbicides - Current Research and Case Studies in Use*. 2013 pages 445-

471. InTech, 2013. http://www.intechopen.com/articles/show/title/herbicide-resistant-weeds-the-technology-and-weed-management.

Qasem, J.R. Recent advances in parasitic weed research, an overview. In: Singh, H.P., Patish, D.R., Kohli, R.K., (eds.), *Weed Management Handbook*. The Haworth Press, Binghamton, NY, USA. 2006a, 627-728.

Qasem, J.R., Foy, C.L. Screening studies on the host range of branched broomrape (*Orobanche ramosa*). *J. Hort. Sci. Biotech,* 2007, 82 (6), 885-892.

Qasem, J.R., Foy, C.L. Weed allelopathy, its ecological impacts and future prospects: A review. *J. Crop Prod.*, 2001, 4, 43-119.

Qasem, J.R., Kasrawi, M.A.Variation of resistance to broomrape (*Orobanche ramosa*) in tomatoes. *Euphytica*, 1995, 81, 109-114.

Quimby, P.C.Jr., Zidack, N.K., Boyette, C.D. A simple method for stabilizing and granulating fungi. *Biocontrol Sci.Technol.,* 1999,9, 5–8. doi: 10.1080/09583159929857.

Raaimakers, D., Raaimakers, J., Borg, S. J. ter, Nassib, A. M., Pieterse, A. H. Effect of sowing date on *Orobanche erenata* infestation in Egypt. *FABIS Newslett.*1988, 22, 33-39.

Racovitza, A. Research on the susceptibility of various species of *Nicotiana* and various cultivars of tobacco to *Orobanche ramosa*. In: Anonymous., *Proceedings, Symposium on Parasitic Weeds, Malta,* 1973. *European Weed Research Council*, Wageningen, 1973 pp. 194-204.

Raju, C.A. Variability in tobacco germplasm towards *Orobanche* infections. In: Moreno, M.T., Cubero, J.I., Berner, D. D., Joel, D., Musselman, L.J., Parker, C. (eds.), *Advances in Parasitic Plant Research. Sixth International Parasitic Weed Symposium*, April 17-18,1996, Cordoba, Spain, pp. 609-614.

Rao, P.G. A rapid method for studying the germination of the root parasitic *Orobanche cernua* Loeff. var. *desertorum* (Beck). *Sci. Cult.*, 1955, 21, 258-262.

Rao, R., Joshi, B.G, Sarma, C.B. Maintenance of *Orobanche cernua* on brinjal plants. *Indian J. Agric. Sci.*, 1982, 52, 764-767.

Razavifar, Z., Karimmojeni, H., Sini, F.G. Effects of wheat-canola intercropping on *Phelipanche aegyptiaca* parasitism. *J. Plant Prot. Res.*, 2017, 57 (3), 268-274.

Reinke, H., Rosenzweig, A., Clausm, K.M., Chisholm, C., Jensen, P. DPX-E 9636, experimental sulfonylurea herbicide for potatoes. In: *Proceedings of the Brighton Crop Protection Conference;*, November 18-21, 1991, Brighton, UK, pp. 445-451.

Rice E.L. *Allelopathy*. Academic Press. New York, 1974.

Rice, E.L. *Allelopathy*. 2nd ed. Academic Press. Orlando. Fl. pp. 422, 1984.

Ristau, R.J. Dodder. Colorado State University Cooperative Extension. 2001, 1995-2001.

Robert, S., Dubreuil, D., Simier, P., Praderre, J.P., Fer, A. Inhibition studies on mannose 6-phosphate reductase purified from *Orobanche ramosa*. *Carbohydr. Lett.*, 1999, 3, 231-238.

Rodri´Guez-Conde, M. F., Moreno, M. T., Cubero, J. I., Rubiales, D. Characterization of the *Orobanche–Medicago truncatula* association for studying early stages of the parasite–host interaction. *Weed Res.*, 2004, 44, 218–223.

Rodriguez-Ojeda, M.I., Alonso, L.C., Fernandez-Escobar, J. Effect of different crops on the germination of *Orobanche cernua* Loelf. (*O. cumana* Wallr) seeds. In: Fer, A., Thalouarn, P., Joel, D.M., Musselman, L.J., Parker, C., Verkleij, J.A.C., (eds.), *Proceedings of the 7th International Parasitic Weed Symposium*, Nantes, France, 2001, 124.

Rubiales, D. Eating broomrape? In: Cuero, J. I., Moreno, M. T., Burials, D., Seller , J. (eds.), *Resistance to Orobanche: The State of the Art*, Junta de Andalucía, Consejería de Agricultura y Pesca, Sevilla, Spain, 1999, pp. 195-199.

Rubiales, D., Fernández-Aparicio, M. Innovations in parasitic weeds management in legume crops. A review. *Agronomy for Sustainable Development, Springer Verlag/EDP Sciences/INRA*, 2012, 32 (2), 433-449.

Rubiales, D., Fernández-Aparicio, M., Vurro, M., Eizenberg, H. Editorial: advances in parasitic weed research. *Front. Plant Sci.*, 2018. 9, 236, 3p. doi: 10.3389/fpls.2018.00236.

Rubiales, D., Flores, F., Emeran, A.A., Kharrat, M., Amri, M., Rojas-Molina, M.M., Sillero, J.C. Identification and multi-environment validation of resistance against broomrapes (*Orobanche crenata* and *Orobanche foetida*) in faba bean (*Vicia faba*). *Field Crops Res.*, 2014, 166, 58–65.

Rubiales, D., Moreno, M.T., Sillero, J.C. Search for resistance to crenate broomrape (*Orobanche crenata* Forsk.) in pea germplasm. *Genet. Resour Crop Evol.*, 2005, 52, 853–861. DOI 10.1007/s10722-003-6116-3.

Rubiales, D., Pujdas, A., Roman, B. Occurrence and distribution of *Phytomyza orobanchia* feeding on broomrapes (*Orobanche* spp.) in southern Spain. In: Fer, A., Thalouarn, P., Joel, D.M., Musselman, L.J., Parker, C., Verkleij, J.A.C., (eds.), *Proceedings of the 7th International Parasitic Weed Symposium,* Nantes, France, 2001, pp. 254-257.

Rubiales, D., Rojas-Molina, M.M., Sillero, J.C. Characterization of resistance mechanisms in faba bean (*Vicia faba*) against broomrape species (*Orobanche* and *Phelipanche* spp.). *Front. Plant Sci.* 2016. DOI:10.3389/fpls.2016.01747.

Rubiales, D., Sillero, J.C., Cubero, J.I. Broomrape (*Orobanche crenata* Forsk.) resistance in peas (*Pisum sativum* L.). In: *3rd European Conference on Grain Legumes. Opportunity for High Quality, Healthy and Added-Value Crops to Meet European Demands*. 14-19 November 1998, Valladolid, Spain.

Rubiales, D., Sillero, J.C., Moreno, M.T. *Orobanche crenata* in winter chickpea. *Grain Legumes*, 1999, 26, 7.

Ruso, J., Melero-Vara, J.M., Dominguez, J., Miller, J., Fernandez-Martinez, J.M. Survey of broomrape (*Orobanche cernua* Loefl.) resistance to collections of cultivated sunflower inbred lines and wild species of *Helianthus*. In: Pieterse, A.H.; Verkleij, J.A.C., ter Borg, S.J., (eds.), *Biology and Management of Orobanche. Proceedings of the Third International Workshop on* Orobanche *and Related* Striga *Research*, November 8-12, 1993. Amsterdam, Netherlands. 1994.

Ruso, J., Sukno, S., Dominguez-Gimenez, J., Melero-Vara, J.M., Fernandez-Matinez, J. Screening of wild *Helianthus* species and derived lines for resistance to several populations of *Orobanche cernua*. *Plant Dis.*, 1996, 80, 1165-1169.

Saber, H.A., Omer, M.A., El-Hady, M.M., Mohmoud, S.A., Abou-Zeid, N.M., Radi, M.M. Performance of a newly-bred faba bean line (X-843) resistant to *Orobanche* in Egypt. In: Kroschel, J., Abderahibi, M., Betz, H. (eds.), *Advances in Parasitic Weed Control at On-Farm Level. Vol. II. Joint Action to Control Orobanche in the WANA Region*. Margraf Verlag, Weikersheim, Germany. 1999, pp. 251-257.

Saghir, A. R. Dormancy and germination of *Orobanche* seeds in relation to control methods. In: ter Borg, S. J., (ed.), *Proceedings Workshop Biol. Control* Orobanche, Wageningen, 1986, pp. 25-34 Wageningen: LH/PVO.

Saghir, A.R. An overview of new techniques in *Orobanche* control. *Plant Prot.Q.*, 1994. 9, 77-80.

Saghir, A.R., Janudi, A., Shafyuddin, M. The use of germination stimulants for the control of *Orobanche* in tomato. In: *10th International Congress of Plant Protection1983. British Crop Protection Council*, 282.

Salem, S. M.; El-Shiny, O.M.A., Ahmed, A.H.H. Comparative studies of the effectiveness of fusilade and glyphosate in controlling *Orobanche crenata* parasitizing *Vicia faba* plants. *Bulletin of Faculty of Agriculture, University of Cairo* 1989, 40, 633-650.

Salle, G., Aber, M., Le-Coz, S. Comparative histo-physiological study of the germination and establishment of *Cuscuta*. In: Parker, C., Musselman, L.J., Polhill, R. M., Wilson, A.K., (eds.), *Proceedings of the Third International Symposium on Parasitic Weeds*. Aleppo, Syria. ICARDA, 1984, pp. 10-21.

Samejima, H., Sugimoto, Y. Recent research progress in combating root parasitic weeds. *Biotechnol. Biotechnol. Equip.*, 2018, 32 (2), 221-240, DOI:10.1080/13102818.2017.1420427.

Sanchez, J., Jurado-Exposito, M., Lopez-Granados, F., Castejon-Munoz, M., Garcia-Torres, L. Pronamide applied to sunflower seeds for *Orobanche cumana* control. *Weed Technol.*, 2003, 17, 314-319.

Sand, P.F. *Orobanche ramosa* in Taxas. In: *Meeting of Weed Science Society of America 1983. Animal and Plant Health Inspection Survey, Plant Protection and Quarantine*, Hyattsville, Maryland, USA. Abstract no. 47.

Sandri, G., Sandri, A., Martini, G. Protection of tobacco against *Orobanche*. *Informatore Agraio*, 1998, 54, 74-75.

Sang, H.C., Yoneyama, K., Takeuchi, Y., Joel, D.M. Fluridone and norflurazon, carotenoid-biosynthesis inhibitors, promote seed conditioning and germination of the holoparasite *Orobanche minor*. *Physiol. Plant.*, 2004, 120, 328–337.

Sarychev, Yu., Trofimova, L.I., Strelyaeva, N.I., Vinogradov, V.A. Phenotypic expression of genes for resistance to broomrape in tobacco. *Genetika* (*Moscow*), 1982, 18, 114-148.

Sauerborn, J. Angiospermous root parasites and fungi as natural antagonists. In: Fer, A., Thalouarn, P., Joel, D.M., Musselman, L.J., Parker, C., Verkleij, J.A.C., (eds.), *Proceedings of the 7th International Parasitic Weed Symposium*, Nantes, France, 2001, pp. 248-251.

Sauerborn, J. *Parasitic Flowering Plants: Ecology and Management*. Weikersheim, Germany: Verlag Josef Margraf. 1991, 129 p.

Sauerborn, J., Buschmann, H., Ghiasi, K.G., Kogel, K.H. Benzothiadiazole activates resistance in sunflower (*Helianthus annuus*) to the root-parasitic weed *Orobanche cumana*. *Phytopathology*, 2002, 92, 59-64.

Sauerborn, J., Saxena, M.C. Effect of soil solarization on *Orobanche* spp. infestation and other pests in faba bean and lentil. In: Weber, H.C., Forstreuter, W., (eds.), *Proceedings of 4th International Symposium on Parasitic Flowering Plants*. Marburg., Germany: Philipps-Universitat. 1987, pp. 733-744.

Sauerborn, K.H., Linke, M.C., Saxena, M.C., Koch, W. Solarization; a physical control method for weed and parasitic plants (*Orobanche* spp.) in Mediterranean agriculture. *Weed Res.*, 1989, 29, 391-397.

Saxena, M.C., Linke, K.H., Saurborn, J. Integrated control of *Orobanche* in cool-season flooding food legumes. In: Pieterse, A.H., Verkleij, J.V.C., Ter Borg, S.J. (eds.), *Biology and Management of Orobanche*, The Netherlands, 1994, pp. 419-431.

Schlingloff, S., Alkamper, J. Competition for moisture between dry peas and weeds at a semi-arid site in Syria. In: *16th German Conference on Weed Biology and Control*, Stuttgart-Hohenheim, Germany, March 10-12 , 1992, *Sonderheft*, 13, 221-228.

Schluter, K., Aber, M. Chemical control of *Orobanche crenata* in field beans in Morocco. *Zeitschrift fur Pflanzenkrankheiten und Pflanzenschutz*, 1980, 87, 433-438.

Schnell, H., Kunisch, M., Saxena, M.C., Sauerborn, J. Simulation of the seed bank dynamics of *Orobanche crenata* Forsk. In some crop rotations common in northern Syria. *Expe. Agric.*, 1996, 32, 395-403.

Schnell, H., Linke, K.H., Sauerborn, J. Trap cropping and its effect on yield and *Orobanche crenata* Forsk. Infestation on following pea (*Pisum sativum* L.) crops. *Trop. Sci.*, 1994, 34, 306-314.

Serghini, K., Perez de Luque, A., Castejon-Munoz, M., Garcia-Torres, L., Jorrin, J. V. Sunflower (*Helianthus annuus* L.) response to broomrape (*Orobanche cernua* Loefl.) parasitism: induced synthesis and excretion of 7-hydroxylated simple coumarins. *J. Exp. Bot.*, 2001, 52, 2227-2234.

Sestacova, T., Gisca, I., Cucereavîi, A., Port, A., Duca, M. Expression of defense-related genes in sunflower infected with broomrape. *Biotechnol Biothchnol Equip.*, 2016, 30, 685–691.

Shabana, Y.M., M¨uller-St¨over, D., Sauerborn, J. Granular pesta formulation of *Fusarium oxysporum* f. sp. *orthoceras* for biological control of sunflower broomrape: efficacy and shelf-life. *Biol. Cont.*, 2003, 26,189–201.

Shi, R., Li, M. The genus *Orobanche* as food and medicine: An ethnopharmacological review. *Ethnopharmacology*, 2020. Available at: https://doi.org/10.1016/j.jep.2020.113154.

Sillero, J.C., Cubero, J.I., Rubiales, D. Resistance to broomrape (*Orobanche crenata*) in *Lathyrus*. In: Fer, A., Thalouarn, P., Joel, D.M., Musselman, L.J., Parker, C., Verkleij, J.A.C., (eds.), *Proceedings of the 7th International Parasitic Weed Symposium*, Nantes, France, 2001a, 224-227.

Sillero, J.C., Moreno, M.T., Rubiales, D. Low induction of *Orobanche crenata* seed germination in wild legume species. In: Fer, A., Thalouarn, P., Joel, D.M., Musselman, L.J., Parker, C., Verkleij, J.A.C., (eds.), *Proceedings of the 7th International Parasitic Weed Symposium,* Nantes, France,2001b., p. 234.

Sillero, J.C., Perez-De-Luque, A., Rubiales, D., Moreno, M.T. Effect of the sowing date on the *Orobanche crenata* infection on susceptible and resistant faba bean and vetch varieties. In: *SEMh Congreso 1999: Sociedad Espanola de Malherbologia*, Actas, Logrono, Spain, Noviembre 23-25,1999. Lerida, Spain; Departamento Hortofruticultura, Botanica i Jardineria, Cordoba, Spain, pp. 401-405.

Silverside, A.J. *Orobanche* L. -Broomrapes-BioInfo (UK). Available at: www-biol. paisley.ac.uk/bioref/Tracheophyta/Orobanche.html. Review. 2002b

Silverside, A.J. *Orobanche minor var. minor,* parasitic on *Medicago lupulina*, Hertfordshire, 2002a. Available at: http://www.bioref.lastdragon.org/Magnoliophytina/Orobanche.html

Simier, P., Delavault, P., Thoiron, S., Veronesi, C., Fer, A., Thalouarn, P. Changes in M6PR expression and enzyme activity in *Orobanche ramosa* during its development on tomato: consequences in mannitol level. In: Fer, A., Thalouarn, P., Joel, D.M., Musselman, L.J., Parker, C., Verkleij, J.A.C., (eds.), *Proceedings of the 7th International Parasitic weed Symposium,* Nants, France, 2001, 147-150.

Singh, S.J., Joshi, H.K. A new record of witches' broom on *Orobanche cernua* parasitising little-leaf diseased eggplant. *International Journal of Tropical Plant Diseases*, 1992, 10, 73-77.

Slavov, S.B., Batchvarova, R.B., Valkov, V.T. Possibilities for obtaining tobacco to *Orobanche* spp. by chemical mutagenesis. In: Fer, A., Thalouarn, P., Joel, D.M., Musselman, L.J., Parker, C., Verkleij, J.A.C., (eds.), *Proceedings of the 7th International Parasitic weed Symposium,* Nants, France, 2001, pp. 88-91.

Smaling, E.M.A., Stein, A., Sloot, P.H.M. A statistical analysis of the influence of *Striga hermonthica* on maize yields in fertilizer trails in southwestern Kenya. *Plant Soil*, 1991, 138,1-8.

Sobrino Vesperinas, E. Search for resistance to *Orobanche ramosa* L. in rapeseed. Cruciferae Newsletter, 1985, No. 10, 120-121.

Solymosi, P. Biology of broomrape (*Orobanche ramosa* L.) and possible control methods. *Novenyvedelem,* 1998, 34, 469-475.

Song, W.J., Cao, D.D., Jin, Z.L., Zhou, W.J. Studies on factors influencing the seed germination of root parasitic plant *Orobanche*. *Seed*, 2005, 24, 44-47.

Song, W.J., Zhou, W.J., Jin, Z.L., Zhang, D., Yoneyama, K., Takeuchi, Y., Joel, D.M. Growth regulators restore germination of *Orobanche* seeds that are conditioned under water stress and suboptimal temperature. *Aust. J. Agr. Res.*, 2006, 57, 1195-1201.

Spelce, D.L., Musselman, L.J. *Orobanche minor* germination with strigol and GR compounds. *Zeitschrift fur Planzenzenphysiologie*, 1981. 104, 281-283.

Spencer, K.A. Notes on the African Agromyzidae. 2. *Deutsche Entomologische Zeitschrift* (DEZ), 1961, 8(5), 415–430.

Spencer, K.A., Agromyzidae (Diptera) of economic importance. *Series Entomologica.* Vol. 9. W. Junk, The Hague, the Netherlands. 1973, pp. 418.

Stewart, G.R., Press, M.C. The physiology and biochemistry of parasitic angiosperms. *Annu. Rev. Plant Physiol.Plant Mol. Biol.*, 1990, 41,127-151.

Strobel, G.A., Kenfield, D., Bunkers, G., Sugawara, F., Clardy, J. Phytotoxins as potential herbicides. *Experientia*, 1991, 47, 819-826.

Sugimoto, Y. Germination stimulants for the seeds of root parasitic weeds. *J. Pestic. Sci.*, 2000, 25, 438-440.

Sukno, S., Ruso, J., Melero-Vara, J., Fernandez-Martinez, J. M. Evaluation of wild *Helianthus* germplasm for resistance to highly virulent races of *Orobanche cernua* Loefl. and transfer to cultivated sunflower. In: Moreno, M.T., Cubero, J.I., Berner, D., Joel, D., Musselman, L.J., Parker, C., (eds.), *Advances in Parasitic Plant research. Sixth International parasitic Weed Symposium,* April 17-18. Cordoba, Spain. 1996, 626-632.

Surov, T., Aviv, D., Aly, R., Joel, D.M., Goldman-Guez, T., Gressel, J. Germination of transgenic asulam-resistant potatoes to facilitate eradication of parasitic broomrapes (*Orobanche* spp.), with the *sul* gene as the selectable marker. *Theoretical and Applied Genetics*, 1998, 96, 132-137.

Syka, E.K., Economou, G. Use of herbicide resistant crops in Greece for control of *Orobanche* and other weeds. In: *Proceedings of the Joint Working Groups and MC meeting of COST Action 849, Broomrape Biology, Control and Management*, September15–17, 2005. Reading University, UK, pp. 35–36. Available at: http://cost849.ba.cnr.it/i.

Syka, E.K., Eleftherohorinos, I..G. *Orobanche ramosa* L. (broomrape) control in tomato (*Lycopersicon esculentum* Mill.) with chlorsulfuron, glyphosate and imazaquin. *Weed Res.*, 1991, 31, 19- 27.

Szökö, G. Investigations on the broomrape-mining fly on sunflower-broomrape. *Mosonmagyaróvári Agrártudomány Föiskola, Közlemények* (Hungary), 1968, 11(1), 137–141.

Takeuchi, Y., Omigawa, Y., Ogasawara, M., Yoneyama, K., Konnai, M., Wosham, A.D. Effects of brassinosteroids on conditioning and germination of clover broomrape (*Orobanche minor*) seeds. *Plant Growth Regul.*,1995, 16, 153-160.

Tawfik, M.F.S., Awadallah, K.T., Shalaby, F.F. Biology of *Phytomyza orobanchia* Kalt. (Diptera; Agromyzidae). *Bulletin de la Société Entomologique d'Egypte,* 1976, 60, 53–64.

Teimouri Jervekani, M., Karimmojeni, H., Brainard, D.C., Jafari, M. Sesame genotype influences growth and phenology of *Phelipanche aegyptiaca. Ann Appl Biol.,* 2016, 169, 46–52.

Thijn, R., Brummelkamp, T.R., Bernards, R., Agami, R. A system for stable expression of short interfering RNAs in mammalian cells. *Science*, 2002, 296, 550–553.

Thomas, H., Heller, A., Sauerborn, J., Muller-Stover, D. *Fusarium oxysporum* f. sp. *orthoceras*, a potential mycoherbicide, parasitize seeds of *Orobanche cumana* (sunflower broomrape): a cytological study. *Ann.Bot.*, 1999b, 83, 453-458.

Thomas, H., Sauerborn, J., Muller-Stover, D., Kroschel, J. Fungi of *Orobanche aegyptiaca* in Nepal with potential as biological control agents. *Biocontrol Sci.Technol.*, 1999a, 9, 379-381.

Thomas, H., Sauerborn, J., Muller-Stover, D., Ziegler, A., Bedi, J. S., Kroschel, J. The potential of *Fusarium oxysporum* f. sp. *orthoceras* as a biological control agent of *Orobanche cumana* in sunflower. *Biol. Control*, 1998, 13, 41-48.

Thorogood, C. J., Rumsey, F. J., Harris, S. A., Hiscock, S. J. Gene flow between alien and native races of the holoparasitic angiosperm *Orobanche minor* (Orobanchaceae). *Plant Syst. Evol.*, 2009, 282, 31–42. DOI 10.1007/s00606-009-0204-6.

Tokasi, S., Bannayan, Aval M., Mashhadi, H.R., Ghanbari, A. Screening of resistance to Egyptian broomrape infection in tomato varieties. *Planta Daninha*, 2014, 32, 109–116.

Tokunaga, T., Hayashi, H., Akiyama, K. Medicaol, a strigolactone identified as a putative didehydro-orobanchol isomer, from *Medicago truncatula. Phytochemistry.*, 2015, 111, 91–97.

Tomilov, A.A., Tomilova, N.B., Wroblewski, T., Michelmore, R., Yoder, J.I. Trans-specific gene silencing between host and parasitic plants. *Plant J.* 2008, 56, 389–397.

Torres, L.G., Granados, F.L. Control of broomrape (*Orobanche crenata* Forsk.) in broad bean (*Vicea faba* L.) with imidazolinones and other herbicides. *Weed Res.*, 1991, 31, 227-235.

Tóth, P., Hudec, K., Cagáň, Ľ. Natural enemies of *Orobanche* species in Slovakia (Tóth). *Joint Working Groups and MC Meeting of COST Action 849, Broomrape Biology, Control and Management,* September15-17, 2005, Reading University, UK. Abstract, pages 28-29.

Trabelsi, I., Abbes, Z., Amri, M., Kharrat, M. Performance of faba bean genotypes with *Orobanche foetida* Poir. and *Orobanche crenata* Forsk. infestation in Tunisia. *Chil. J. Agric. Res.*, 2015, 75, 27–34.

Trabelsi, I., Abbes, Z., Amri, M., Kharrat, M. Study of some resistance mechanisms to *Orobanche* spp. infestation in faba bean (*Vicia faba* L.) breeding lines in Tunisia. *Plant Prod. Sci.*, 2016, 19, 562–573.

Traoré, D., Vincent, C., Stewart, R.K. *Smicronyx guineanus* Voss and *S. umbrinus* Hustache (Coleoptera: Curculionidae): potential biocontrol agents of *Striga hermonthica* (Del.) Benth. (Scrophulariaceae). In: Hess, D.E, Lenné, J.M., (eds.). *Report on the ICRISAT Sector Review for Striga Control in Sorghum and Millet. International Crops Research Institute for the Semi-Arid Tropics*, Bamako, Mali, 1999, pp. 105-115.

Trenchev, G. The possibility of using *Phytomyza orobanchia* Kalt. for the control of broomrape. *Rasteniev'D. Nauki,* 1981,18, 112-119.

Tsybul'skaya, G.N., Degtyarev, B.G., Fedoryak, N.A., Skoklyuk, A.N. Determination of viability of *Phytomyza puparia. Zashchita Rastenii*, 1978, 5, 29-30.

Tsybul'skaya, G.N., Skoklyuk, A.N. Calculating release of *Phytomyza. Zashchita Rastenii*, 1978, 11, 49.

University of California, Division of Agricultural Sciences, California Agricultural Experimental Station, Extention Service. Branched broomrape (*Orobanche ramosa*) a threat to California crops 1965.

V"lchev, H., Bakhariev, D., Masheva, S., Toskov, K. Possibilities for the use of chemicals in the control of *Orobanche* (*Orobanche ramosa*) in tomato stands. *Rasteniev'dni Nauki*, 1995, 32, 62-64.

Vachev, T., Ivanova, D., Minkov, I., Tsagris, M., Gozmanova, M. Trafficking of the Potato spindle tuber viroid between tomato and *Orobanche ramosa. Virology,* 2010, 399, 187–193.

Vail, S.L., Dailey, O.D. Jr., Connik, W.J., Pepperman, A.B. Strigol synthesis and related structure-bioactivity studies. In: *The Chemistry of Allelopathy. Biochemical Interactions among Plants*, Thompson, A., Ed., Washington, D.C., USA, *American Chemical Society Symposium Series*, 1985, 268, 445-456.

Valkov, V.T., Bachvarova, R.B., Slavov, S.B., Atanassova, S.I., Atanassov, A.I. Genetic transformation of tobacco for resistance to Basta. *Bulg. J. Agric. Sci.*,1998, 4, 1-7.

Van Hezewijk, M. J., Pieterse, A. H., Saxena, M. C., Borg, S. J. ter. Relationship between sowing date and *Orobanche* (broomrape) development on faba bean (*Vicia faba* L.) and lentil (*Lens culinaris* Medikus) in Syria. In: Weber, H. C., Forstreuter, W., (eds.), *Proceedings. 4th Internat. Syrup. Parasitic Flowering Plants*, Marburg: *Philipps Universitait Marburg, Germany,* 1987pp. 377-390.

Van Hezewijk, M. J., Verkleu, J.A.C. The effect of nitrogenous compounds on *in vitro* germination of *Orobanche crenata* Forsk. *Weed Res.*, 1996, 36(5), 395-404.

Veronesi, C., Bonnin, E., Calvez, S., Thalouarn, P., Simier, P. Activity of secreted cell wall-modifying enzymes and expression of peroxidase-encoding gene following germination of *Orobanche ramosa. Biol. Plant.*, 2007, 51 (2), 391-394.

Veronesi, C., Delavault, P., Simier, P. Acibenzolar-S-methyl induces resistance in oilseed rape (*Brassica napus* L.) against branched broomrape (*Orobanche ramosa* L.). *Crop Prot.*, 2009, 28, 104–108.

Vinogradov, V. A., Mironov, E. K., Strelyaeva, N. I., Sarychev, Y. F. Breeding tobacco for resistance to *Orobanche. Tabak,* 1981, 4, 55-57.

Vinogradov, V.A., Mironov, E.K., Sarychev, Yu. F. Degree of field resistance to *Orobanche* in tobacco mutants, hybrid and varieties. *Plant Breeding Abstracts*, 1984, 54, 3464.

Vorländer, C. *Biological Control of Orobanche crenata in Food Legumes*. ICARDA, *Progress Report 1*. 1988, Aleppo, Syria

Vouzounis, N. Lessons learned from integrated control of *Orobanche* in Cyprus (Vouzounis). *Joint Working Groups and MC meeting of COST Action 849, Broomrape Biology, Control and Management,* September 15-17, 2005, Reading University, UK.

Vouzounis, N.A., Americanos, P.G. Control of *Orobanche* (broomrape) in tomato and eggplant. *Technical Bulletin 196.Agricultural Research Institute, Ministry of Agriculture, Natural Resources and the Environment*, Nicosia, Cyprus. 1998.

Vranceanu, A.V., Soare, G., Craiciu, D., Voinescu, G. Adaptability of some Romanian sunflower hybrids in Europe. *Analele Institutului de Cercetari pentru Ceeale si Plante Tehnice, Fundulea*, 1990, 58, 61-80

Vranceanu, A.V., Tudor, V.A., Stoenescu, F.M., Pirvu, N. Virulence group of *O. cumana* Wallr., differential hosts and resistance sources and genes in sunflower. In: *Proceedings of the Ninth International Sunflower Conference,* Torremolinos, Spain. 1980, 74–82.

Wakabayashi, T., Joseph, B., Yasumoto, S., Akashi, T., Aoki, T., Harada, K., Muranaka, S., Bamba, T., Fukusaki, E., Takeuchi, Y., Yoneyama, K., Muranaka, T., Sugimoto, Y., Okazawa, A. Planteose as a storage carbohydrate required for early stage of germination of *Orobanche minor* and its metabolism as a possible target for selective control. *J. Exp. Bot.*, 2015, 66, 3085–3097.

Watzl, O. Die kleine Sommerwurz Fliege (*Phytomyza orobanchia* Kaltenbach). *Neuheiten auf dem Gebiete des Pflanzenschutzes,* 1936, 29(6), 228–230.

Wegmann, K. Progress in *Orobanche* research during the past decade. In: Wegmann K., Musselman L.J., Joel. D.M., (eds.), *Current Problems of Orobanche Rresearch, Proceedings of the 4th International Orobanche Workshop*, Sept. 23-26, 1998, *Institute for Wheat and Sunflower "Dobroudja",* Albena, Bulgaria, 1998 pp. 13-17.

Wegmann, K., von Elert, E., Harloff, H.J., Stadler, M. Tolerance and resistance to *Orobanche*. In: Wegmann, K., Musselman, L.J., (eds.), *Progress in Orobanche Research. Proceedings, International Workshop on* Orobanche *Research, Obermarchtal,* 1989, Eberhard-Karls-Universitat, Tubingen, 1991, 318-321.

Westwood, J.H., dePamphilis, C.W., Das, M., Fernández-Aparicio, M., Honaas, L.A., Timko, M.P., Wafula, E.K., Wickett, N.J., Yoder, J.I. The parasitic plant genome project: new tools for understanding the biology of *Orobanche* and *Striga*. *Weed Sci.*, 2012, 60, 295–306.

Westwood, J.H., Foy, C.L. Influence of nitrogen on germination and early development of broomrape (*Orobanche* spp.). *Weed Sci.*, 1999, 47, 2-7.

Wigchert, S.C.M., Kuiper, E., Boelhouwer, G.J., Nefkense, G.H.L., Verkleij, J.A., Zwanenburg, B. Dose-response of seeds of the parasitic weeds *Striga* and *Orobanche* toward the synthetic germination stimulants GR 24 and Nijmegen 1. *J. Agric. Food Chem.*, 1999, 47(4), 1705-1710.

Willem, J., Thuring, J.W.J.F., Heinsman, N.W.J.T., Jacobs, R.W.A.W.M., Nefkens, G.H.L., Zwanenburg, B. Asymmetric synthesis of all stereoisomers of demethylsorgolactone. Dependence of the stimulatory activity of *Striga hermonthica* and *Orobanche crenata* seed germination on the absolute configuration. *J. Agric. Food Chem.*, 1997, 45, 507-513.

Worsham, A.D., Klingman, G.C., Moreland, D.E. Promotion of germination of *Striga asiatica* seed by coumarin derivatives and effect of seedling development. *Nature*, 1962, 195, 199-200

WSSA. Orobanche-The Broomrapes. Available at: https://wssa.net/wp-content/themes/WSSA/WorldOfWeeds/orobanche.html (Accessed 12 March 2021)

Xie, X., Yoneyama, K., Harada, Y., Fusegi, N., Yamada, Y., Ito, S., Yokota, T., Takeuchi, Y., Yoneyama, K. Fabacyl acetate, a germination stimulant for root parasitic plants from *Pisum sativum*. *Phytochemistry*, 2009, 70, 211–215.

Xie, X., Yoneyama, K., Kusumoto, D., Yamada, Y., Yokota, T., Takeuchi, Y., Yoneyama, K. Isolation and identification of alectrol as (þ)-orobanchyl acetate, a germination stimulant for root parasitic plants. *Phytochemistry*, 2008, 69, 427–431.

Xie, X., Yoneyama, K., Yoneyama, K. The strigolactone story. *Annu. Rev. Phytopathol.,* 2010, 48, 93–117.doi:10.1146/annurev-phyto-073009-114453.

Yadav, A., Malik, R.K., Punia, S.S., Malik, R.S., Kumar, R., Yadav, K.K., Singh, S. Broomrape (*Orobanche aegyptiaca* Pers.) infestation in oilseed-rapes and management options. In: *Technical Bulletin* (6), Department of Agronomy and Directorate of Extension Education, Chaudhary Charan Singh Haryana Agricultural University, Hisar. 2005, pp. 18.

Ye, X., Chen, J., McErlean, C.S.P., Zhang, M., Yu, R., Ma, Y. The potential of foxtail millet as a trap crop for sunflower broomrape. *Acta Physiol Plant,* 2017, 39, 1 Available at: https://doi.org/10.1007/s11738-016-2300-x

Yokota, T., Sakai, H., Okuno, K., Yoneyama, K., Takeuchi, Y. Alectrol and orobanchol, germination stimulants for *Orobanche minor*, from its host red clover. *Phytochemistry*, 1998,49, 1967-1973.

Yoneyama, K., Ogasawara, M., Takeuchi, Y., Konnal, M., Sugimoto, Y., Sassa, T., Yoshida, S. Effect of jasmonates and related compounds on seed germination of *Orobanche minor* Smith and *Striga hermonthica* (Del.) Benth. *Biosci. Biotechnol. Biochem.*, 1998a, 62, 1448-1450.

Yoneyama, K., Takeuchi, Y., Ogasawara, M., Konnal, M., Sugimoto, Y., Sassa, T. Cotylenins and fusicoccins stimulate seed germination of *Striga hermonthica* (Del.) Benth and *Orobanche minor* Smith. *J. Agric. Food Chem.*, 1998b, 46, 1583-1586.

Yoneyama, K., Takeuchi, Y., Yokota, T. Natural germination stimulants for *Orobanche minor*. In: Fer, A., Thalouarn, P., Joel, D.M., Musselman, L.J., Parker, C., Verkleij, J.A.C., (eds.), *Proceedings of the 7th International Parasitic Weed Symposium*, Nantes, France, June 2001, p. 123.

Yoneyama, K., Xie, X., Sekimoto, H., Takeuchi, Y., Ogasawara, S., Akiyama, K., Hayashi, H., Yoneyama, K. Strigolactones, host recognition signals for root parasitic plants and arbuscular mycorrhizal fungi, from Fabaceae plants. *New Phytol.*, 2008, 179, 484–494.

Yoneyama, K., Xie, X., Yoneyama, K., Takeuchi, Y. Strigolactones: structures and biological activities. *Pest Manag Sci.*, 2009, 65, 467–470.

Yoshida, S., Cui, S., Ichihashi, Y., Shirasu, K. The haustorium, a specialized invasive organ in parasitic plants. *Ann. Rev. Plant Biol.*, 2016, 67, 643–667.

Zahran, M.K. Efficiency of terbutol in broomrape control and broadbean improvement. *Pest Article and News Summaries (PANS)*, 1973, 19, 230-232.

Zahran, M.K. Weed and *Orobanche* control in Egypt. In: Hawtin, G., Webb, C., (eds.), *Faba Bean Improvement.* Tha Hague, Netherlands Martinus Nijhoff, 1982, pp. 191-198.

Zain, A.E.M. *Benzothiadiazole Influencing Host-Parasite Interactions.* M.Sc. Thesis, University of Hohenheim, Stuttgart, Germany. 2002.

Zaitoun, F.M.F., Al-Menoufi, O.A., Weber, H.C. Susceptibility of five varieties of *Vicia faba* to infection with *Orobanche crenata*. In: Ransom J.K.; Musselman L.J.; Worsham A.D. , Parker, C. (eds.), *Proceedings of the 5th International Symposium of Parasitic Weeds*, Nairobi, Kenya, 24-30 June 1991, pp. 208-216.

Zaitoun, F.M.F., Sabra, F.S. Resistance of the new Egyptian faba bean cultivars against *Orobanche crenata.* In: Americanos, P.G., Abu- Irmaileh, B.A., Saghir, A.R., (eds.), *Proceedings of the First Near East Conference on Improved Weed Management*, February, 2000, Khartoum, Sudan, 2000, pp. 225-230.

Zaitoun, F.M.F., Saleh, M.E., Borg, S.J.Ter. Effect of some trace elements on germination and seedling development of *Orobanche.* In: Brown, H., Cussans, G.W., Devine, M.D., Duke, S.O., Fernandez-Quintanilla, C., Helweg, A., Labrada, R.E., Landes, M., Kudsk, P., Streibig, J.C. (eds.),*Proceedings of the Second International Weed Control Congress*, June 25-28, 1996, Copenhagen, Denmark, volumes 1-4.

Zaki, M.A., El-Metwaly, H.S., Hassan, R.A. Studies on dodder (*Cuscuta* spp.) control. In: Maillet J., (ed.), *Comptes-rendus 6eme Symposium Mediterranean EWRS*, Montpellier, France, 13-15 mai 1998, pp. 147-150.

Zehar, N., Fer, A. Germination of *Orobanche ramosa* is controlled by gibberellins and ethylene. In: Fer, A., Thalouarn, P., Joel, D.M., Musselman, L.J., Parker, C., Verkleij, J.A.C., (eds.), *Proceedings of the 7th International Parasitic Weed Symposium*, Nantes, France, 2001, p. 122.

Zehhar, N., Ingouff, M., Bouya, D., Fer, A. Possible involvement of gibberellins and ethylene in *Orobanche ramosa* germination. *Weed Res.*, 2002, 42, 464-469.

Zemrag, A., Bajja, M. Characterization of *Orobanche* spp. in Morocco and the effect of some trap crops on *Orobanche crenata* Forsk in faba bean (*Vicia faba* L.). In: Fer, A., Thalouarn, P., Joel, D.M., Musselman, L.J., Parker, C., Verkleij, J.A.C., (eds.), *Proceedings of the 7th International Parasitic Weed Symposium*, June 2001, Nantes, France, 2001 p. 300.

Zerman, N. Parasitic weeds of economic importance in North Africa: problems and management. In: Americanos, P.G., Abu- Irmaileh, B.A., Saghir, A.R. (eds.), *Proceedings of the First Near East Conference on Improved Weed Management*, February, 2000. Khartoum, Sudan, pp. 257- 260.

Zonno, M.C., Vurro, M. Effect of fungal toxins on germination of *Striga hermonthica* seeds. *Weed Res.*, 1999, 39, 15-20.

Zonno, M.C., Vurro, M. Inhibition of germination of *Orobanche ramosa* seeds by *Fusarium* toxins. *Phytoparasitica*, 2002, 30, 519-524.

Zwanenburg, B., Mwakaboko, A.S., Reizelman, A. structure and function of natural and synthetic signaling molecules in parasitic weed germination. *Pest Manag Sci.*, 2009, 65, 5, 478–491. Available at: https:doi.org/10.1002/ps.1706.

Zwanenburg, B., Pospíšil, T. Structure and activity of strigolactones: new plant hormones with a rich future. *Mol. Plant*, 2013, 6 (1), 38–62.

Zwanenburg, B., Reizelman, A. En route to the isolation and characterization of the strigolactone receptor using biotin labeled strigolactone analogues. In: Fer, A., Thalouarn, P., Joel, D.M., Musselman, L.J., Parker, C., Verkleij, J.A.C., (eds.), *Proceedings of the 7th International Parasitic Weed Symposium*, Nantes, France, 2001, pp. 102-105.

Websites

html; http://www-biol.pai

http://www.dijon.inra.fr/malherbo/hyppa/hyppa-a/oracr_ah.htm (cited by Qasem, 2006a)
http://www.uni-hohenheim.de/~www380/parasite/oro_path.htm (cited by Qasem, 2006a)
http://www.wssa.net/subpages/weed/larrymitich/Orobanche.html

http://www-aes.tamu.edu/mary/brmrape/brmrape.htm (cited by Qasem, 2006a) http://www-biol.paisley.ac.uk/bioref/Plantae/Orobanche.html (cited by Qasem, 2006a) https://wssa.net/wp-content/themes/WSSA/WorldOfWeeds/orobanche.html

CHAPTER 2

Desert Hyacinth (*Cistanche* spp.)

Jamal Ragheb Said Qasem

INTRODUCTION

Cistanche spp. are annuals or perennials, phanerogamic plants, holo-root-parasites attack host plant species in arid and semi-arid land. *Cistanche* contains 27 species in the world (www.theplantlist.org), distributed in the tropical and sub-tropical regions in different parts of the world across Eurasia and North Africa, China, Iran and India (Piwowarczyk *et al.*, 2016). These parasites belong to the family Orobanchaceae to which other parasitic species belong to this family and follows the suborder Lamialis which in turn belongs to the class Eudicots and the phyla Angiosperms. Among the commonly known *Cistanche* species are *tubulosa, lutea, salsa, ambigua, deserticola, phelypaea, sinensis, armena* and *violacea.* Their heights ranged between 30-65 cm or may be more in some cases in the desert. These parasites attack plant species in dry regions naturally higher and mostly shrubs and trees. Although different species of *Cistanche* are known but *tubulosa* and *lutea* species are the main representatives of the genera although sometimes considered as synonyms. *Cistanche* spp. parasitize species of trees, shrubs and perennial herbs of different plant families including Chenopodiaceae, Zygophyllaceae, Tamaricaceae and Plumbaginaceae. Chenopodiacea species are the most commonly attacked including species of *Atreplix, Salsola, Haloxylon, Salicornia, Suaeda* and *Anabasis.* Certain species were also found parasitizing *Tamarix* spp., *Casuarina anguistifolia* and *Punica granatum* (Qasem, 2011).

Cistanche spp. are not able to produce their own food or do photosynthesis because they all lack chlorophyll pigments and thus can't do photosynthesis and are completely dependent on host species for food, water and mineral nutrients through an absorptive organ "Haustorium' that inserted in host tissues, tightly connected to the host vascular tissues and may cause death to host plants under heavy infestation. However, other pigments are present such as yellow, red and pink colours based on the species attacked.

Chemical analysis of these parasites revealed the presence of many natural chemicals of pharmacological importance making them of value and widely used in folk medicine to treat different ailments. *Cistanche deserticola* Y. C. Ma and *Cistanche tubulosa* (Schrenk) Wight are widely used in medicine. Although *Cistanche* spp. are obligate parasites since devoid of chlorophyll pigments but certain species were found growing by themselves in the fields, or host root exudates, extracts or a mixture of extract and growth regulators failed to induce their germination under laboratory conditions. However, flouridone has been repeatedly reported to enhance seed germination of *C. tubulosa* by affecting the GA/ABA balance. Studies on the control of these parasites are scanty but may be controlled by different methods including hand pulling or hoeing just at emergence and before start flowering, treatment by very low rates of certain translocated herbicides such as glyphosate and 2,4-D, and may be by biological control using certain insects such as *Eumerus* bulb flies.

GERMINATION AND DEVELOPMENT

Information on germination and growth of *Cistanche* spp. are generally limited because these parasites were not given enough emphasis from researches since are not parasitizing important economic food crops. However, *Cistanche* spp. reproduce from seeds that are produced in high numbers and fall on the soil upon maturation or disperse into other regions in different methods and through agents from their production sites. Seeds are somehow like *Orobanche* seeds in morphology and requirements for germination but appear of larger size (Fig. 1.2.1). Seeds start germination when conditions required for germination, including moisture and temperature, are available. The optimal temperature for the seeds germination of three species of *Cistanche* was at 20-30°C, and the seeds did not germinate at sub-or supraoptimal temperatures (5 and 35°C). Seed dormancy was observed for *C. tubulosa*. Seeds won't germinate without exposure to natural germination stimulants released from the roots of a host plant (Li *et al.*, 1989; Cheni *et al.*, 2011). Cold stratification breaks seed dormancy in *C. deserticola* through a mechanism that mainly relies

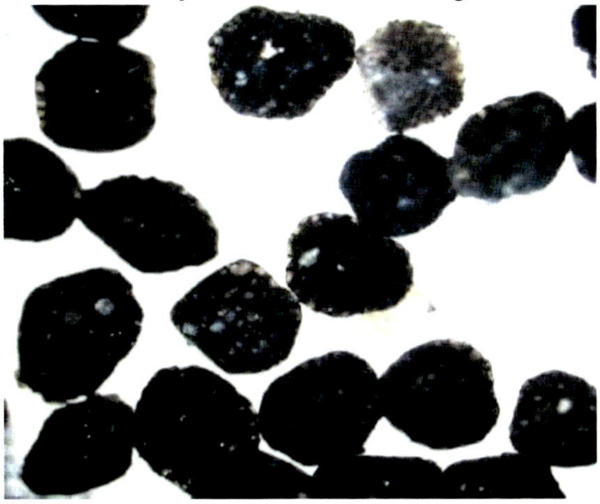

Fig. 1.2.1. Seeds of *Cistanche salsa*

on modifying the ABA contents in the seeds and gibberellins (GAs) synthesis occurs widely during seed conditioning (Suzuki *et al.*, 1994; Takeuchi *et al.*, 1995; Zehhar *et al.*, 2002; Song *et al.*, 2005), GAs and ABA play key roles for the seed germination of *C. tubulosa* while fluridone inhibited ABA biosynthesis but increased the concentration of GAs in seeds, and thus fluridone might initiate other processes associated with germination (Chen *et al.*, 2016).The seed may stay in the soil for many years until the host is available and starts secreting chemical stimulants from host roots that enhance parasite seed germination. However, seeds of *Cistanche violacea* and *Cistanche phelypaea* species showed no response when exposed to these stimulants in the field and away from host plants. In addition, a mixture of host root extract and growth regulators failed to stimulate most parasite seed germination.

Germinated seed produces the germination tube that resembles root but is not a root, this extends toward the host root system. The germination tube forms an absorptive tuberous organ on its terminal portion that acts to penetrate host tissues reaching its vascular bundles connecting itself to.

CONTACT AND ATTACHMENT

Seedling of *Cistanche* lacks the radicle, hypocotyl and cotyledons. Instead, the germinating embryo develops a tube-like organ that contact the host root. The formation of tuberous shape organ, sucker and the bud are the main steps toward parasite development. A tuberous shape organ grow toward the host root and when became in contact with root surface it is swollen forming the sucker (haustorium) that strongly penetrates the root epiderms and vascular bundles releasing brown-color materials. The embryonic haustorium is formed with 6 or more small leaves, the tuberous organ break down and the seed coat cracked. Because of parasitism, the host root side from the tuberous stem is swollen and the other side disintegrates and disappears. Once the parasite reaches the central core of root xylem of the host plant, the haustorium replaces the metaxylem cells and ensures physical support and supply of resources. The parasite starts growing after connect itself to the root system of its host and then gradually developed as a result of absorbing the required sucking water (Fig. 1.2.2.), food and mineral nutrients through an absorptive organs (Haustoria) necessary for its growth from the host species and thus starts forming a fleshy thick stem that developed from the tuberous absorptive organ connected to the host vascular bundles. This stem develops underground in the late summer or autumn.

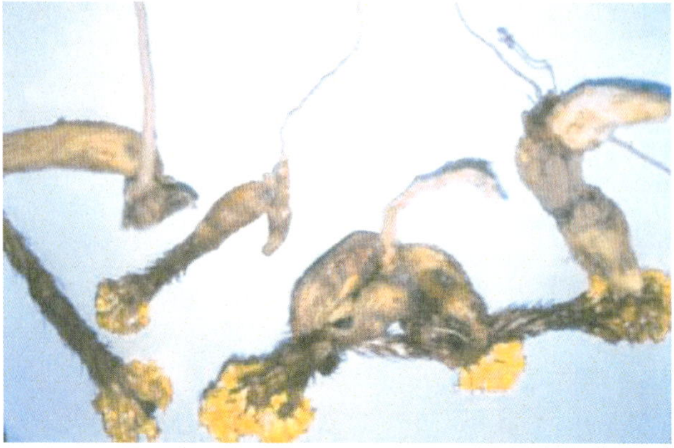

Fig. 1.2.2. The *Cistanche* tuberous growth and its developed stem on roots of *Suaeda* sp.

The parasite spend most of the period required for its growth below the soil surface, It remains dormant until spring or summer then projects above the soil as unbranched thick stem in the different species. The stem carries no true leaves but rather fleshy disintegrated enclosing the stem. The flowering stem apex is opened giving flowers of white, yellow, pink or light red in colors according to the species. Flowers are large compared with those of *Orobanche*. Parasite seeds are produced in a huge number and disperse in different ways by wind, water, animals, agricultural machines, labors, polluted seeds and through packed agricultural materials such as peatmoss, organic fertilizers and others. After seed set and release, the inflorescence first withers (usually within 2-3 weeks), followed by the underground stem.

BIOLOGY, ECOLOGY AND PHYSIOLOGY

Representative Species

Cistanche tubulosa (Schenk) Wight

Synonym is *Cistanche lutea* (Desf.) Hoftmgg. and Link. Common name is Desert Hyacinth. According to Karim (1978), *C. tubulosa* is described as a root parasite, annual or perennial, 30-65 cm long, fleshy herb, non-green (commonly lacking chlorophyll) brownish. The subterranean stem is thick, bulb-like, and fleshy without root cap, 10-16 cm long and 2-2.5 cm diameter. The lower part of the erect plant is leafless, flowerless, thick simple, in upper part leaves present alternate, scale-like, 3-4.5 cm long and 1.5- 2.5 cm wide. The floral part 1/3 to ½ of the height of the erect plant. Inflorescence is a raceme; flowers 4-4.5 cm long. Compact, in dense spike of 16-38 or more, flowers sub-sessile, solitary, irregular, gamopetalous, asymmetric, bisexual, subtended by a large, thick, brownish, glabrous, ovate-lanceolate bract, $\pm$ 2.5 cm long and $\pm$ 1.0 cm wide, branches linear-lanceolate shorter than calyx $\pm$ 2.0 cm long $\pm$ 0.5 cm wide. Calyx 5-lobed, united below, tubular-campanulate, slightly toothed, straightened, lobe ovate elliptic 1.5-2.0 cm long and $\pm$ 1.0 cm wide.

Cistanche deserticola Y. C. M

This is a desert inhabiting root parasite plant. It parasitizes the root of *Haloxylon ammodendron*, whose woody stem and branches are good firewood called "coal in the desert" and used by the local people and building materials and whose tender branches are camel feed in the winter and spring. If the host continues to be used to the present degree and is exterminated, then this valuable medicinal herb is feared to disappear also from the harsh desert (Peng Hua and XU Zaifu, undated).

The relationship between the parasitic *Cistanche* spp. and their hosts is a negative parasitism. The host is loosing while the parasite is the beneficial species, grows and increases and the negative relationship is extenuated with high infestation and reduction in growth factors necessary in the habitats at which these are growing which mostly arid or semi-arid regions, saline and poor in most cases. However, in spite the lack or absence of any estimation for the losses caused by these parasites but severe infestation of trees and shrubs under such conditions result in weakness at a great extent. On the other hand, perennial species such as *Alhagi*, *Centaurea*, and *Prosopis* may be completely killed in low or moderate parasite infestation. May be some tree species such as *Casuarina* appears less influenced by the parasites especially at low infestation.

DISTRIBUTION AND HOST RANGE

As indicated earlier *Cistanche* spp. are distributed in warm regions in the world including tropical and sub-tropical desert in the southern regions of Spain, Portugal, North Africa, China, Eurasia, Armenia, Iran, India, Mongolia (Shahi and Saeidi, 2010; Piwowarczyk *et al.*, 2016), Syria, Iraq, Turkey, Jordan, Palestine, Saudi Arabia and Gulf states (Table 1.2.1). They are mainly parasites on roots of plants in the family Chenopodiaceae (Moreno Moral *et al.* 2018).

Table 1.2.1. *Cistanche* species and their distribution

Species	Distribution	Reference
Cistanche deserticola Y. C. M	China, Mongolia	http://yahoo.com/winghop-fung/cisthanche.html
Cistanche lanzhouensis Zhi Y.Zhang	China	Zhang, 1984
Cistanche laxiflora Aitch. and Hemsl.	Iran	Mehrvarz and Shavvon, 2008
Cistanche lutea (Desf.) Hoffmgg. Link)	and Jordan	Qasem, 2009
Cistanche phelypaea (L.) Cout.	Spain, Pakistan, Iraq, Arabian Gulf States	Fawaz and Musselman, 1999
Cistanche salsa (C. A. Meyer) G. Beck	Jordan	Qasem, 2009
Cistanche tubulosa Wight	Pakistan, Iraq, Arabian Gulf, States, Spain, Jordan, Palestine.	Holm *et al*, 1979; Karim, 1978; Al-Khesraji *et al*, 1987

Host species of different *Cistanche* species are shown in Table 1. 2. 2. *Cistanche deserticola* lives parasitically on the roots of psammophyte *Haloxylon ammodendron* (Chenopodiaceae) and *C.*

tubulosa is a parasite on the roots of *Tamarix*, which are best for precautions sand species. *Cistanche salsa*, a plant parasite found on the roots of shrub plants, such as Chenopodiaceae, Tamaricaceae, Zygophyllaceae, which are used for improving saline land (Fu *et al*., 2018). *Cistanche armena* parasitizing shrubs from the genera *Salsola* L. (Chenopodiaceae) and *Alhagi* Gagnebin (Fabaceae) (Piwowarczyk *et al.,* 2017). *Cistanche tubulosa* is an obligate parasite of the roots of the perennial plant *Tamarix chinensis.*

ECONOMIC IMPORTANCE

Cistanche salsa

This parasitic species contains volatile oils, non-volatile phenylethanoid glycosides, iridoids, lignans, alditols, oligosaccharides, a monoterpene and polysaccharides. The essential oil includes alkanes, alcohols, aldehydes, and some heterocyclic compounds. In addition, acetylacteoside, acteoside, tubuloside, isoacteoside, echinacoside, cistanosides, alpha and beta-unsaturated ketone of phenl-2-propenoyl, phenylethanoids, and phenylpropanoid glycosides are also present. The plant has antioxidative activity, neuroprotective activity in human cells, has an anti-osteoporotic effect, provides antibody production in human lymph node lymphocytes, is antinociceptive and has anti-inflammatory properties. *Cistanche salsa* extracts enhance antibody production in isolated human lymph node lymphocytes *in vitro.* The plant is the main source of the drug cistanche. Echinacosides are an immune booster, immune regulator, and a remedy for colds and flu. *Cistanche* has antioxidant abilities and could also help soften the effects of ageing. It is applied as a tonic and/or in a formula for chronic renal disease, impotence, female infertility, morbid leucorrhoea, profuse metrorrhagia, and senile constipation. *Cistanche salsa* extract can accelerate the proliferation of fibroblasts and promotes the production of neurons by accelerating the growth of the neuritis. It also has properties including the prevention of damage caused by cerebral ischemia-reperfusion; protection against apoptosis of the CA1 region of the hippocampus; and increasing the number of neurotransmitters such as dopamine, noradrenaline, and serotonin (5-HT) in the rat brain. The acteoside-rich fraction of C. salsa extract can inhibit reactive oxygen species (ROS), prevent DNA damage, enhance superoxide dismutase activity and prevent lipid peroxidation. Acteoside has an extremely strong antioxidative effect that is 15 times stronger than resveratrol and five times stronger than vitamin C. *Cistanche salsa* extract can activate lymphoid cells and increases the killed rate of cancer cells. The monoterpene is an anti-osteoporotic compound. A wild flower and parasite. It is widely used for firewood.

Table 1. 2. 2. Host range of certain *Cistanche* species.

Species	Recorded host species	Reference
Cistanche armena Medik.	*Alhagi maurorum* Medik. *Halostachys belangeriana* (Moq.) Botsch. *Salsola dendroides*	Piwowarczyk *et al.*, 2017 Aghababyan, 2013.
Cistanche deserticola Y. C. M	*Haloxylon ammodendron* C.A. Mey Bunge	http://yahoo.com/wingh op-fung/cisthanche.html; Wu *et al.*, 2019.
Cistanche lanzhouensis Zhi Y. Zhang	*Nitraria, Haloxylon*	Zhang, 1984.
Cistanche laxiflora Aitch. and Hemsl.	*Pytropyrum, Seidlitzia*	Mehrvarz and Shavvon, 2008.
Cistanche phelypaea (L.) Cout.	*Hammada elegans* (Bunge) Botsch. M	Fahmy *et al.*, 1996.
Cistanche salsa (C. A. Meyer) G. Beck	*Atriplex* spp. *Salsola* spp. *Anabasis* spp.	Qasem, 2009, Mehrvarz and Shavvon, 2008.
Cistanche tubulosa (Schrenk) Hook	*Anabasis articulate* (Forssk) Moq. *Arthrocnemum macrostachyum* (Moric.) C.Koch *Atriplex* spp. *Salsola* spp. *Haloxylon salicornicum* (Moq.) Bung. *Capparis spinosa* L. *Zygophyllum coccinum* L. *Salvadora* spp. *Calotropis* spp. *Calligonum* sp. *Prosopis* spp., *Pteropyrum* sp. *Tamarix* spp.	Fahmy *et al.*, 1996. Qasem, 2009 Karim, 1978. Baquar and Tasnif, 1967 Mehrvarz and Shavvon, 2008.

Cistanche deserticola

It is known as "Herba Cistanche", and used to treat several ailments. This herbal medicine consists of the pulpy stem with scales of *Cistanche deserticola*. It is characterized as sweet and salty in flavour, warm in property, acting on the kidney and large intestine channels. Its effect is reported to be tonifying the kidney, replenishing the vital essence and blood, and moistening the bowels to relieve constipation (Qasem, 2020). In a laboratory study on mice, *Cistanche deserticola* extracts appeared to enhance the swimming capacity of mice by decreasing muscle damage, delaying the accumulation of lactic acid and by improving energy storage.

Due to its high potential as herbal medicine *Cistanche deserticola* is at present included in the list of highly endangered plant species.

Cistanche tubulosa

The plant contains alkaloids, coumarins, saponins, apigenin, terpenes, tannins, acteoside, echinacoside, cistanoside, campneoside II, phenylethanoid glycosides, iridoid glycosides, apigenin glycoside, monoterpenoid adoxosidic acid, 8-epiloganic acid, geniposidic acid, mussaenosidic acid, (+)-syringaresinol-O-□-D-glucopyranoside, 8-hydroxygeraniol, mussaenosidic acid and (+)-syringaresinol-O-□-D-glucopyranoside. In the dry bulbs, □-sitosterol, D-mannitol, daucosterol, succinic acid, D-glucose and D-fructose are present (Qasem, 2020).

The plant is used medicinally to prevent senile dementia, and inhibit the aggregation of blood platelets. It is believed to be important in the treatment of male impotence, seminal emission, the general lassitude of the loins and knees and constipation. None of these, however, are supported by scientific evidence. High doses of the chemical compound acteoside, isolated from *Cistanche tubulosa*, affected biomarkers related to ageing in mice. It may delay ageing, probably because of the antioxidant action since it enhances immunity in ageing mice. Anti-fatigue properties have been shown in mice. In addition, *C. tubulosa* extract exhibits hypocholesterolemic activity in diet-induced hypercholesterolemia mice. It appeared to decrease muscle damage and delay the accumulation of lactic acid. Externally, the plant is used as an astringent for wounds and in curing porous scarfskin or skin infected by psoriasis. It enhances the sex function, available for essence and blood, and is commonly used in the treatment of male impotence, infertility (both sexes), metrorrhagia, vaginal discharge, powerlessness in bones and muscles, depleted blood, jaundice, constipation, improving renal function, enhancing memory, anti-senile dementia, regulating the immune function, acting on large intestine channels, kidney tonic, replenishing vital essences and blood, and moistening the bowels to relieve constipation. It is used for diarrhoea and skin sores (Qasem, 2020).

In folk medicine, the plant is widely consumed as a precious medicinal herb in traditional Arab and Chinese medical practices for the treatment of kidney deficiencies, female infertility, morbid leucorrhoea, neurasthenia, and senile constipation induced by colonic inertia. Some other evidence for the medicinal importance of different *Cistanch*e species has also been reported. The polysaccharides isolated from *Cistanche deserticola* have *in vitro* immunomodulatory effects in high doses, promoting the phagocytic and secretory functions of the phagocytic ability of peritoneal macrophage. *Cistanche salsa* extracts enhance antibody production in isolated human lymph node lymphocytes *in vitro*. Anti-fatigue properties have also been shown in mice. It was reported that high doses of *Cistanche herba* extract significantly enhanced learning and memory, as demonstrated by the passive avoidance test and novel object recognition test.

The plant has always been used as food by people for nourishing purposes. The thick fleshy stem is boiled and eaten, and newly emerged young stems are used in food as asparagus substitutes. The plant's powder is added to meat dishes as a spice. The plant is a wild flower and parasite. However, it has been mentioned in a few references as being poisonous.

CONTROL

As such *Cistanche* is not a plant destined for destruction. Instead it is labeled as one of the most valuable endangered species (Hue and Zaifu, undated). However, for certain less desirable species research has been underway to develop means of control but in general there is clearly a lack of literature on the control of these parasites since these species were not fully studied because appear in the desert, tropical and sub-tropical regions parasitizing wild grown or rangeland plant species of less importance than crop plants and thus the topic is not much attractive or of priority to researchers as well as to farmers. These species however, have not been generally reported to attack species of economic importance or cultivated food crops which may be due to shortage in the information on these species or lack of published research work on hosts and their ability to parasitize crop plants if conditions permit. However, Qasem (2009) reported that *C. salsa* parasitize pomegranate trees and found connected to the root system of this species while *C. tubulosa* was found on *Casuarina equisetifolia*. Control of these species may become necessary once found on economic crops through the following methods:

1. Hand pulling or hoeing or through tillage once they appear above the soil level and before start flowering and seeding. Parasite shoots can be gathered and burned. In addition, collecting these species for public use in folk medicine may be sufficient for the control of these parasites and placing the species as endangered possibly disappearing soon from the environment.

2. Chemical control using herbicides at very low rates of application and thus avoid their harmful effects on treated host plants. It may possible using 2,4-D and glyphosate or other systemic herbicides against these parasites through injection in the host stem.
Flouridon was found to enhance germination of *C. tubulosa* under laboratory conditions. It regulates seed conditioning and germination in many parasitic angiosperms. The percentage of *Cistanche tubulosa* seed that germinate can be significantly increased by flouridon treatment (Wang *et al.*, 2006). Flouridon is known to act as an abscisic acid (ABA)-biosynthesis inhibitor, yet the underlying mechanisms of how it affects seed conditioning and germination have not been fully disclosed. *Cistanche tubulosa* seeds treated with flouridon will germinate even in the absence of a host. Therefore it may be used as germination stimulant in absence of host plants but under local conditions these parasites do not cause danger effects on economic crops that require such a control treatment.

3. Biological control methods using parasites natural enemies. Some species of *Cistanche* are attacked at flowering stage and some insects feed on their seeds. In Safawa (southern desert of Iraq), Al-Khazraji and Abdel Wahid (1987) reported insect pests attacking parasitic flowering plants. The insect was identified to be *Eumerus* bulb flies (Syrphidae). The larvae were found feeding on *Cistanche* and then transferred to *Orobanche*, as well as to *Cynomorium*. There were two species identified as large type and small type. According to the authors, both are symmetrical, have the same development time, and the nature of their feeding is the same. There was high infestation rate and sharing by both larvae on both plants. The larvae of some species (*Eumerus ammophilus* Paramonov, *E. arnoldii* Stackelberg, *E. cistanchei* Efflatoun, *E. compertus* Villeneuve, *E. mucidus* Bezzi (Stackelberg 1961; Shaumar and Kamal, 1978)) have been found in the tubers and shoots of holoparasitic plants of *Cistanche phelypaea* (L.) Cout. (as *Cistanche tinctoria* (Forssk.) G. Beck), or *C. lutea* (Desf.) Hoffmgg. and Link), *C. tubulosa* (Schenk) Hook. fil., or *Cistanche violacea* (Desf.) Hoffmgg. and Link, in Iraq and Egypt (Waitzbauer, 1976; Shaumar and Kamal, 1978; Al-Khezraji *et al.,* 1987). *Cistanche armena* (K. Koch) M.V. Agab. was mentioned to be attacked by the larvae of *Eumerus mucidus* (Diptera: Syrphidae). Larvae was reported for the first time as boring channels in the fresh tubers and shoots of the holoparasitic plant *Cistanche armena* (K. Koch) M.V. Agab., causing damage and breakage (Piwowarczyk, and Mielczarek, 2018). Unidentified thrip was also found attacking *Cistanche tubulosa* seeds and appeared widely spread in Al-Azraq area in Jordan (Qasem, 2011), the larvae was found feeding on seeds in the capsules. In addition, the beetle *Epicometis squalida* (*Tropinota squalida* Scopoli, 1783) adults (Fig. 1.2.3) were found on the floral parts of *Cistanche tubulosa* in Al-Azraq region

in Jordan and appeared stunted and destructive to the parasite flowers. The beetle belongs to the family Scarabaeidae, subfamily Cetoniinae. The larvae were reported to feed on roots, while the adults can be encountered from May through July feeding on flowers of different plant species.

Fig. 1.2.3. The beetle (*Epicometis squalida*) adults on the flowers of *C. tubulosa*

DESERT HYACINTH (*CISTANCHE* SPP.) IN JORDAN

GENERAL BACKGROUND

Cistanche spp. are well-known desert plants able to grow in dry and saline conditions in Jordan. Three species were found and appeared largely confined to forage species, mainly belonging to the Chenopodiaceae, and possibly to certain fruit and forest trees (Table 1.2.3). *Cistanche tubulosa* was found parasitizing *Casuarina equisetifolia*, *Punica granatum*, and *Trifolium pratense*. It was growing in a close proximity to *Eucalyptus camaldulensis* and *Opuntia ficus-indica* plants, although the actual attachments to these probable hosts were not checked by excavations. Otherwise, all the *Cistanche* species attacked mainly shrubs and forage plants, including some noxious perennial weeds, such as *Prosopis farcta*, *Zilla spinosa*, *Centaurea postii* and *Alhagi maurorum*, in the desert and arid rangelands. To our knowledge, this is the first report of *Cistanche* spp. parasitizing certain species including: *Haloxylon persicum* by *Cistanche lutea*; *A. maurorum*, *C. equisetifolia*, *Centaurea postii*, *P. farcta*, *P. granatum*, *Salicornia herbacea*, *Salsola vermiculata*, *Suaeda asphaltica*, *Suaeda vera*, *Tamarix pentandra*, and *T. pratense* by *C. tubulosa*; and *Achillea* sp., *Atriplex halimus*, *H. persicum*, *Haloxylon salicornicum*, *Salsola* spp., *Suaeda* sp., and *Z. spinosa* by *Cistanche salsa*. Illustrations (plates 1-73) showing *Cistanche* spp. in Jordan with some of their hosts are presented. Although the survey study covered most locations in the country, other novel parasitic associations might have been missed or not recorded.

Cistanche (3 species) attacking 23 species of 7 families (connection with 5 species was not checked).

C. tubulosa 19 host species (connection with 5 species was not checked)
C. lutea 1 species
C. salsa 10 species (connection with 2 species was not checked)
Most parasitized families: Chenopodiaceae (7 species), Leguminosae (3 species).

Table 1.2.3. Cistanche species, hosts and their families, status and intensity of infection found in Jordan.

Parasitic species/host	Family	Status	Intensity of infection	Prevalence
***a. Cistanche tubulosa* (Schrenk) Hook**				
Acacia cyanophylla L. (NC)	Leguminosae	C	Light	R
Alhagi maurorum Medik.	Leguminosae	W	High	L
Atriplex glauca L.	Chenopodiaceae	W	High	L
Atriplex halimus L.	Chenopodiaceae	W	High	VC
Centaurea postii Boiss.	Compositae	W	Moderate	L
Casuarina equisetifolia L. ex J.R. and G. Forst	Casuarinaceae	C	Light	L
Eucalyptus camaldulensis Dehnh. (NC)	Myrtaceae	C	Moderate	R
Olea europaea L.(NC)	Oleaceae	C	Light	R
*Opuntia ficus-indica*L. (NC)	Cactaceae	C	Moderate	R
Pinus sp. (NC)	Pinaceae	C	Moderate	R
Prosopis farcta (Banks and Sol.) Macbride	Leguminosae	W	Moderate	L
Punica granatum L.	Punicaceae	C	Light	R
Salicornia herbacea L.	Chenopodiaceae	W	Light	L
Salsola vermiculata L.	Chenopodiaceae	W	Light	CO
Suaeda asphaltica Boiss.	Chenopodiaceae	W	Moderate	CO
Suaeda vera Forsk.	Chenopodiaceae	W	Light	L
Tamarix pentandra Pall.	Tamaricaceae	W	Light	R
Trifolium pratense L.	Leguminosae	C	Light	CO
***b. Cistanche lutea* (Desf.) Hoffmanns and Link**				
Haloxylon persicum Bunge ex Boiss. and Buhse	Chenopodiaceae	W	Light	L
Olea europaea L. (NC)	Oleaceae	C	Light	R
***c. Cistanche salsa* (C. A. Meyer) G. Beck**				
Achillea sp.	Compositae	W	Moderate	L
Aeluropus lagopoides (Linn.) Trin. Ex Thw.	Gramineae	W	Light	R
Anabasis sp.	Chenopodiaceae	W	Moderate	CO
Atriplex halimus L.	Chenopodiaceae	W	Moderate	CO
Haloxylon persicum Bunge ex Boiss. and Buhse	Chenopodiaceae	W	High	VC
Haloxylon salicornicum (Moq.) Bunge ex Boiss.	Chenopodiaceae	W	Moderate	VC
Salsola sp.	Chenopodiaceae	W	Moderate	CO
Suaeda sp.	Chenopodiaceae	W	Moderate	L
Zilla spinosa (L.) Prantl	Cruciferae	W	Light	L

NC = Attachment was not confirmed
R= rare that only on few plants in 1-2 sites of a biogeographical region
W: wild, C: cultivated
L = Limited: on many plants localized in certain locations of 1 or 2 biogeographical regions
CO = Common on certain plant species in > one biogeographical region
VC = very common found on many plant species in different locations of different biogeographical regions
Source: modified after Qasem, 2009

SOME ILLUSTRATIONS ON *CISTANCHE* SPP. AND THEIR HOSTS IN JORDAN

Cistanche tubulosa

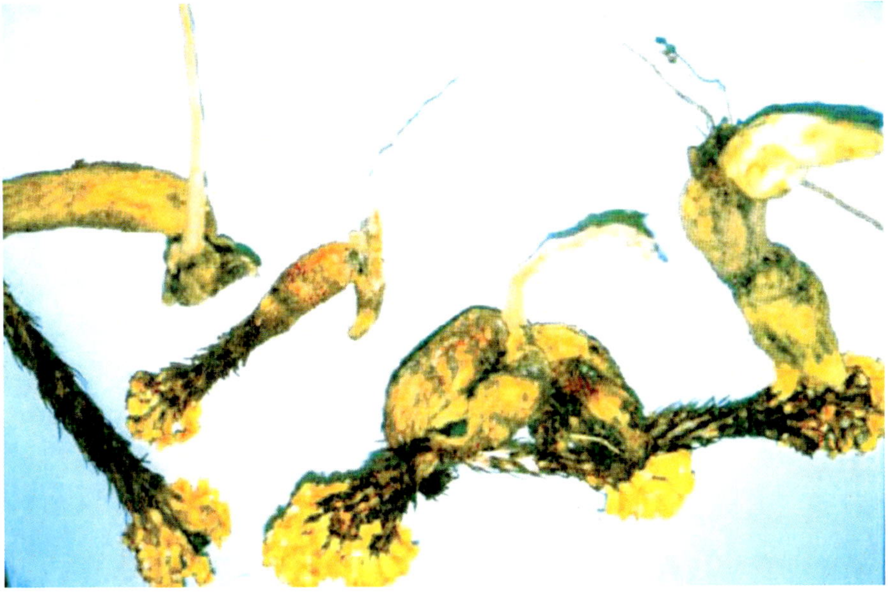

Plate 1. *Cistanche tubulosa* on *Suaeda* sp

Plate 2. *Cistanche tubulosa* connected with *Suaeda* sp. roots

Plate 3. *Cistanche tubulosa* shoots connected with the same point on host roots showing tuberous lower shoot parts

Plate 4. *Cistanche tubulosa* flowers morphology

Plate 5. *Cistanche tubulosa* at early emergence on *Suaeda* sp.

Plate 6. *Cistanche tubulosa* on *Alḥagi maurorum*

Plate 7. *Cistanche tubulosa* killed *Alhagi maurorum* plants

Plate 8. *Cistanche tubulosa* destruction to *Alhagi maurorum* but the parasite grown on re-vegetated perennial host plants

Plate 9. *Cistanche tubulosa* growing in close to *Cupressus sempervirens*

Plate 10. *Cistanche tubulosa* on *Atriplex halimus*

Plate 11. *Cistanche tubulosa* killing *Atriplex halimus*

Plate 12. *Cistanche tubulosa* on *Atriplex leucoclada*

Plate 13. *Cistanche tubulosa* on *Salvia* sp.

Plate 14. *Cistanche tubulosa* on *Salsola vermiculata*

Plate 15. *Cistanche tubulosa* on *Centaurea* sp.

Plate 16. *Cistanche tubulosa* on *Suaeda* sp.

Plate 17. *Cistanche tubulosa* growing close to *Suaeda* sp.

Plate 18. *Cistanche tubulosa* killed *Zilla spinosa*

Plate 19. *Cistanche tubulosa* on *Polygonum* sp.

Plate 20. *Cistanche tubulosa* growing close to *Arthrocnemum macrostachyum*

Plate 21. *Cistanche tubulosa* on *Achillea* sp.

Plate 22. *Cistanche tubulosa* close to *Peganum harmala*

Plate 23. *Cistanche tubulosa* on *Prosopis farcta*

Plate 24. *Cistanche tubulosa* killing *Prosopis farcta*

Plate 25. *Cistanche tubulosa* close to grapes (*Vitis vinifera*)

Plate 26. *Cistanche tubulosa* close to olive (*Olea europaea*) tree

Plate 27. *Cistanche tubulosa* under *Pinus* sp.

Plate 28. *Cistanche tubulosa* appears near and within pine (*Pinus* sp.) trees

Plate 29. *Cistanche tubulosa* under *Eucalyptus camaldulensis*

Plate 30. *Cistanche tubulosa* killing *Salsola vermiculata* under *Eucalyptus camaldulensis* trees

Plate 31. *Cistanche tubulosa* on *Salsola vermiculata*

Plate 32. *Cistanche tubulosa* on *Casuarina equisetifolia*

Plate 33. *Cistanche tubulosa* on *Tamarix pentandra* and *Atriplex* sp.

Plate 34. *Cistanche tubulosa* on *Acacia cyanophylla*

Plate 35. *Cistanche tubulosa* on *Haloxylon persicum*

Plate 36. *Cistanche tubulosa* on a young *Haloxylon persicum* plant

Plate 37. *Cistanche tubulosa* shoots at mature stage on *Haloxylon persicum*

Plate 38. *Cistanche tubulosa* killed *Haloxylon persicum*

Plate 39. *Cistanche tubulosa* on *Salicornia herbacea*

Plate 40. *Cistanche tubulosa* on *Capparis* sp.

Plate 41. *Cistanche tubulosa* on *Suaeda* sp.

Plate 42. *Epicometis squalida* beetle feeding inside *Cistanche tubulosa* flowers parasitizing *Atriplex* sp.

Cistanche lutea

Plate 43. *Cistanche lutea at emergence*

Plate 44. *Cistanche lutea* on *Haloxylon persicum*

Plate 45. *Cistanche lutea* on *Haloxylon persicum*

Plate 46. *Cistanche lutea* on *Haloxylon salicornicum*

Plate 47. *Cistanche lutea* under olive (*Olea europaea*) trees

Plate 48. *Cistanche lutea* flowers morphology

Cistanche salsa

Early after emergence

Emergence

Fully emerged

Started flowers opening

Death

Plate 49. *Cistanche salsa* at different growth developments

Plate 50. *Cistanche salsa* flowering stem

Plate 51. *Cistanche salsa* on pomegranate (*Punica granatum*)

Plate 52. *Cistanche salsa* close to apricot (*Prunus armeniaca*) tree

Plate 53. *Cistanche salsa* on *Atriplex leucoclada*

Plate 54. *Cistanche salsa* on *Anabasis syriaca*

Plate 55. *Cistanche salsa* on *Atriplex* sp. and close to olive tree

Plate 56. *Cistanche salsa* on *Haloxylon persicum*

Plate 57. *Cistanche salsa* on *Salsola* sp.(upper) and *Erucaria* sp. (lower)

Plate 58. *Cisctanche* sp. on *Anabasis syriaca*

Plate 59. *Cistanche salsa* on *Ferula sainaica*

Plate 60. *Cistanche salsa* on *Achillea* sp.

Plate 61. *Cistanche salsa* on *Salicornia herbacea*

Plate 62. *Cistanche salsa* connected on *Salsola* sp. roots

Plate 63. *Cistanche salsa* on *Salsola* sp.

Plate 64. *Cistanche salsa* on *Achillea* sp.

Plate 65. *Cistanche salsa* on *Haloxylon salicornicum*

Plate 66. *Cistanche salsa* on *Suaeda* sp

Plate 67. *Cistanche salsa* on *Matthiola* sp.

Plate 68. *Cistanche salsa* on a flowering *Matthiola* sp.

Plate 69. *Cistanche salsa* on *Haloxylon persicum*

late 70. *Cistanche salsa* emerged inside *Haloxylon salicornicum* plant

Plate 71. *Cistanche salsa* emerged far distance from *Haloxylon salicornicum* plant

Plate 72. *Cistanche salsa* on *Salsola* sp.

REFERENCES

Aghababyan, M. V. *Flora Mediterranea*, 2013, 23, 131.

Al-Khesraji, T.O., Abdel Wahib, A.U., Annou, M.R. Insect pests attacking parasitic flowering plants in the southern desert of Iraq. *Iraqi J. Agr. Sci. "Zanco"*, 1987, 5, 197–216.

Baquar, S. R., Tasnif, M. *Medicinal Plants of Southern West Pakistan*," Vol. 3, *Pakistan Council of Scientific and Industrial Research Bulletin/Monograph, Karachi*, 1967, p. 56.

Chen, J., Xue, Q. H., McErlean, C. S. P., Zhi, J. H., Ma, Y. Q., Jia, X. T., Zhang, M., Ye, X. X. Biocontrol potential of the antagonistic microorganism *Streptomyces enissocaesilis* against *Orobanche cumana. BioControl.*, 2016, 61,781–791. DOI 10.1007/s10526-016-9738-z.

Chen, Q.L., Jia, Y.M., Wang, Z.F., Shan, C.G., Zhu, J.B., Guo, Y.H. Postembryonic development of *Cistanche tubulosa* (Schrenk) Whigt. *Pak. J. Bot.*, 2011, 43(4), 1823-1830.

Fahmy, G.M., El-Rantawy, H., El Ghani, M.M.A. Distribution, host range and biomass of two species of *Cistanche* and *Orobanche cernua* parasitizing the roots of some Egyptian xerophytes. *J. Arid Environ.*, 1996, 34, 263-276.

Fawaz, A. M., Musselman, L.J. *Cistanche phelypaea*, a native root parasite attacking introduced shrubs. Haustorium, Official Organ of the International Parasitic Seed Plant Research Group 33: *Parasitic Plants Newsletter*, 1999, 35.

Fu, Z., Fan, X., Wang, X., Gao, X. Cistanches Herba: An overview of its chemistry, pharmacology, and pharmacokinetics property. *J. Ethnopharmacol.*, 2018, 219, 233–247).

Holm, L., Pancho, J., Herberger, J., Plucknett, D. *A Geographical Atlas of World Weeds*. Wiley-Interscience, New York, 1979, 440 pp.

Karim, F.M. *Flowering Parasitic Plants of Iraq. Ministry of Agriculture and Agrarian Reform*, Republic of Iraq. 1978.

Li, T.R., Xu, Y.Y., Ge, J.X., Xu, M.Y. The seed germination of *Cistanche deserticola* Ma and its relationship with host plant *Haloxylon ammodendron* Bunge. *Acta Sci. Nat. Univ. Neimongol*, 1989, 20, 395-400.

Mehrvarz, S.S., Shavvon, R.S. Notes on the genus *Cistanche* (Orobanchaceae) in Iran. *Iran J. Bot.*, 2008, 14(2), 95-99.

Moreno Moral, G., Sánchez Pedraja, Ó., Piwowarczyk, R. Contributions to the knowledge of *Cistanche* (Orobanchaceae) in the Western Palearctic. *Phyton (Horn, Austria)*, 2018, 57, 19–36.

Peng, H., Zaifu, XU. (Undated). The threatened wild plants used for medicine as Chinese medicinal herbs. *Xishuangbanna, Chinese Academy of Sciences, Yunnan.* http://monkey.ioz.ac.cn/bwg-cciced/english/bwg-cciced/tech-33.htm.

Piwowarczyk, R., Carlón, L., Kasińska, J., Tofil, S., Furmańczyk, P. Micromorphological intraspecific differentiation of nectar guides and landing platform for pollinators in the Iberian parasitic plant *Cistanche phelypaea* (Orobanchaceae). *Bot. Lett.*, 2016, 163, 47–55.

Piwowarczyk, R., Kwolek, D., Góralski, G., Denysenko, M., Joachimiak, A.J., Aleksanyan, A. First report of the holoparasitic flowering plant *Cistanche armena* on Caspian manna (*Alhagi maurorum*) in Armenia. *Plant Dis.*, 2017, 101, 512. doi.org/10.1094/PDIS-10-16-1469-PDN.

Piwowarczyk, R., Mielczarek, L. First Report of *Eumerus mucidus* (Diptera: Syrphidae) on *Cistanche armena* (Orobanchaceae) and from Armenia. *Fla. Entomol.*, 2018, 101(3), 519-521 URL: https://doi.org/10.1653/024.101.0314.

Qasem, J.R. Parasitic weeds of the Orobanchaceae family and their natural hosts in Jordan. *Weed Biol. Manag.*, 2009, 9 (2), 112-122.

Shahi Shavvon, R., Mehrvarz, S. Pollen and seed morphology of the genus *Cistanche* (Orobanchaceae) in Iran. *Biologia,* 2010, 65 (4), 615–620.
Shaumar, N., Kamal, S. The Syrphidae of Egypt. *Bulletin Mensuel de la Société Linnéenne de Lyon,* 1978, 47, 79–84.
Song, W.J., Cao, D.D., Jin, Z.L., Zhou, W.J. Studies on factors influencing the seed germination of root parasitic plant *Orobanche. Seed*, 2005, 24, 44-47.
Stackelberg, A, A. Palaearctic species of the genus *Eumerus* Mg. (Diptera, Syrphidae). *Trudy Vsesoyuznogo Entomologicheskogo Obshchestva*, 1961, 48, 181–229. (In Russian).
Suzuki, Y., Murofushi, N., Zhang, Y.H., Takeuchi, Y. Endogenous gibberellins in clover broomrape, parasitic plant, and its host, clover. Dependence of the parasite on the host for gibberellins production. *Plant Growth Regul.*, 1994, 13, 63-67.
Takeuchi, Y., Omigawa, Y., Ogasawara, M., Yoneyama, K., Konnai, M., Wosham, A.D. Effects of brassinosteroids on conditioning and germination of clover broomrape (*Orobanche minor*) seeds. *Plant Growth Regul.*, 1995, 16, 153-160.
Waitzbauer, W. *Eumerus compertus* Villeneuve (Dipt., Syrphidae); Larve und Puparium. *Zool. Anz.*, 1976, 196, 16–22.
Wang, H.L., Guo, Y.H., Zhai, Z.X., Chen, Q.L., Yang, T.X., Zhang, J.X. The effect of fluridone on *Cistanche tubulosa* seed germination. *China. J. Chin. Mat. Med.*, 2006, 31, 1638–1639.
Wu, C.-J., Chien, M.-Y., Lin, N.-H., Lin, Y.-C., Chen, W.-Y., Chen, C.-H., Tzen, J. T. C. Echinacoside isolated from *Cistanche tubulosa* putatively stimulates growth hormone secretion via activation of the ghrelin receptor. *Molecules*, 2019, 24, 720, 1-11, doi:10.3390/molecules24040720.
Zehhar, N., Ingouff, M., Bouya, D., Fer, A. Possible involvement of gibberellins and ethylene in *Orobanche ramosa* germination. *Weed Res.*, 2002, 42, 464-469.

Websites
fung/cisthanche.html
http://yahoo.com/winghop-www. theplantlist.org

SECTION II

2.THE FAMILY: SANTALACEAE

Parasitic Weeds of Jordan, 2022, *Part 1*, 272-302

CHAPTER 3

Osyris (*Osyris* spp.)

Jamal Ragheb Said Qasem

INTRODUCTION

Osyris (*Osyris alba* L.) is a woody shrub of 0.2-6m in height depending on the attacked host species, food supply and ecological factors, and turns out constituted from ram celli black and thin on which they dull alternate of dark green color and sessile long narrow leaves 3-8 cm.

Osyris belongs to the Santalaceae family, a dioecious hemiparasite, taking advantage of other plant species (Southern Illinois University Carbondale (SIUC). 2005), its roots are attached to host roots from which they absorb nutrients through relatively large size haustoria. It is widely spread in southern Europe, North Africa, and southwest Asia, mainly in the Mediterranean region (Aronne *et al.*, 1993, Qasem, 2020a). It has been reported as a temperate to tropical, cosmopolitan, except in cold regions (Watson and Dallwitz, 1992). The parasite is frost-sensitive and prefers sandy soils. Flowers are yellow, with three petals and flowering scented (honey odor) in spring during May-July. Pollination is a prerequisite for fruit and seed development. The single seed, containing a small embryo and a large, rich endosperm, is enclosed in a red color fleshy bird-dispersed berry fruit. On maturity, seeds become dark brown. Drupes are non-edible. The parasite has been mentioned to attack different plant species, including almond, grapes, olives, pine, oak, cypress (Qasem, 2006; 2020a) and reported on the edges of pears (Joel *et al*, 1991). Other workers reported *Osyris alba* to infest grapes (Nickrent, 2002; Nickrent and Musselman, 2004).

GERMINATION AND DEVELOPMENT

Osyris spp. are facultative hemiparasites, do not require hosts to complete their life cycle since they are photosynthetic (*e.g. Osyris alba*). Therefore, they are not in need of host root stimulants to start seed germination. They reproduce by seeds that disperse by birds similar to mistletoes seeds or drop on the soil after ripen and the seed is later released. However, since the seed coat is hard, seeds pass through the birds' guts and their germination was mentioned as improved. May be seeds need mechanical abrasion to ease their germination. *Osyris lanceolata* grows from seeds that must be sown fresh. No pretreatment is needed but nicking the base of the seed improves germination. Seed germination can reach 60% after 6 weeks, but there is a general lack of information on seed germination. Many of the seeds are empty or not viable. Seeds of *Osyris* have underdeveloped linear embryos and thus potentially have some dormancy period. Germination temperature is 25-40 °C and high humidity is required. Seeds start germination after 21 days from sowing while germination may be extended to 60 days under ideal conditions. Germination percentage is between 60 to 80.

CONTACT AND ATTACHMENT

After seed germination, the young seedling starts growing and extending its root to the roots of nearby plants and became in contact with the roots of a suitable host. The small parasite seedlings are ready for pricking out 2 to 4 weeks after germination when they have 6-8 leaves and are about 6 cm tall. When found with host roots, they invariably form haustorial connections. These parasites start extracting water and dissolved minerals from host plants (Fig. 2.1.1).

Fig. 2.1.1. *Osyris alba* seedlings emerged from *Retama raetam* root

They can be found in several root-parasitic families such as Olacaceae, Opiliaceae, Santalaceae, Krameriaceae, and Scrophulariaceae (Nickrent, 2002).

BIOLOGY, ECOLOGY AND PHYSIOLOGY

Osyris alba L. (Syn. *Osyris mediterranea* Bubani), a Mediterranean basin shrub or sub-shrub of 0.2 - 6 m and may be more in height. It is a hemiparasite on other plant species and exhibits self-parasitism (Planchon, 1858; Aronne *et. al.*, 1993). *Osyris* belongs to the Santalaceae family that includes almost 30 genera, all are believed facultative parasites. These parasites obtain food and water through absorptive organs (haustoria) attached to the root system of host plants. The parasite is frost-sensitive and prefers sandy soils. *Osyris alba* is of average growth with foliage persisting, coriaceous, alternate, hairless and green, very branched, twigs erect, tapered, spindly and angular. It has woody, erect, or sometimes creeping dark green or brown stem and turns out constituted from ram celli black and thin on which they dull alternate of dark green color and sessile long narrow leaves green clearly, lengthened 3-8 cm (Southern Illinois University Carbondale (SIUC), 2005). Flowers are small dioecious, with three petals and flowering in the spring during May-July, the females are final and solitary and the males are inside bunches. Pollination is required for fruit and seed development which mostly occurs by unspecialized flies and beetles; however, all flowers liberate a nice sweet odour. The number of male flowers is large and they are produced over a long period than the females and provide pollen, nectar and staminal hairs to attract pollinators. The presence and function of stamina hairs with tip cells in *Osyris alba* has been reported. Female flowers producing neither mature pollen, nectar, nor staminal hairs but possess three modified yellow indehiscent anthers containing no viable pollens, which may provide a strong visual feeding stimulus for pollinators. It is suggested that pollinators are attracted by deceit to female flowers by mimicry of the males, and the floral mimicry is, therefore, intraspecific and intersexual. The floral characteristics and flowering phenology of male and female plants are consistent with this kind of mimicry. The female flower possesses a tricarpellary ovary with three ovules, of which the only one develops.

Fruits are drupes globules, red, and non-edible (Fig. 2.1.2). The single seed, containing a small embryo and a large, rich endosperm, is enclosed in a red color fleshy bird-dispersed berry or drupe fruit (Aronne *et al.*, 1993). On maturity, seeds become dark brown. The reduction in seed number

Fig. 2.1.2. *Osyris alba* fruits

per flower to one highly nutrient-invested seed, together with a reduction of the multiflowered inflorescence to a solitary flower and the sequential production of ripe fruits over an extended fruiting season, suggest that the female function is markedly resource-limited. It is suggested that the reproductive characteristics and hemi-parasitism present in *Osyris alba* represent an adaptive survival strategy of the parasite when resources became limited in the parasite habitat.

Osyris lanceolata: This is a sub-shrub, long-lived, hermaphrodite radicular (40-150 cm). It has a green, very angular stem. Leaves are alternate, simple, linear-lanceolate-acute, coriaceous, un inervate and dark green (Watson and Dallwitz,1992). Flowers are yellowish, scented (with honey odor), dioecious, males in small side bunches, solitary females with the extremity unbroken into leaf branch; perianth 3-many; 1 style. Fruit is a drupe, red with maturity. This species is mainly found in arid places, under growth. It flowers from April to June.

***Osyris quadripartita*:** It is an evergreen root hemiparasitic dioecious shrub up to 3 m in height, occurring in a variety of Mediterranean-type scrublands, ranging from the Canary Islands, northwest Africa, and the Iberian Peninsula (Herrera, 1988a). *Osyris quadripartita* parasitizes evergreen and summer drought-deciduous shrubs of many species. The flowering period lasts for nearly 6 months for females and almost the entire year for males, with a peak in May-June for both sexes. The interval between anthesis and fruit ripening is quite variable (13-54 weeks), due to within plant variance in developmental rates. As a consequence, ripe fruits are produced throughout the year, with a major peak in winter and a minor one in the spring (Herrera, 1984; 1985).

Males and females are similar in their proportional use of host-plant species at two study sites and differ significantly at others providing no evidence for niche differentiation of the sexes of *O. quadripartita* along an environmental gradient defined by abiotic factors (moisture, soil nutrients), but favour the view of small - scale segregation based on differential utilization of host–plants (Herrera, 1988b).

DISTRIBUTION AND HOST RANGE

Osyris is widely spread in southern Europe, North Africa, and southwest Asia, mainly in the Mediterranean region (Aronne *et al.*, 1993; Southern Illinois University Carbondale (SIUC), 2005; Qasem, 2020a) and has been reported as a temperate to tropical, cosmopolitan, except in cold regions (Watson and Dallwitz, 1992).

Osyris alba has been old mentioned as parasitizing different perennial plant species of both herbaceous or woody plants of different families of dicotyledons, including *Ailanthus, Rhus coriaria, Ulmus campestris, Jasminum fruticans, Pinus halepensis, Rosa canina, Silene italica, Lychnis dioica* and *Rubia peregrina* (Planchon, 1858), later on other host species have been added as attacked by this parasite including *Hedera helix* (Benharrat *et al.*, 1987), *Amygdalus, Quercus* and *Cupressus* (Joel *et al.*, 1991). It was mentioned also growing on the edges of *Pyrus communis* and *Vitis vinifera* (Nickrent, 2002; Nickrent and Musselman, 2004). However, in more recent studies, Qasem (2006, 2020a) recorded 23 host species 5 of which the parasite attachment on their roots was not possible to confirm but was growing in their vicinity. Host species reported for *Osyris lanceolata* were *Rhus natalensis, Dodonaea viscosa, Tecomaria capensis, Catha edulis, Apodytes dimidiata, Brachystigia spiciformis* and *Cassuarina equisetifolia* (Mwang'ingo *et al.*, 2005).

ECONOMIC IMPORTANCE

Since *Osyris* spp. are facultative hemiparasites that do photosynthesis and can live separately, the suitability of a host species may also be influenced by competition for light between the parasite and its host plants. Hemiparasites strongly compete with the host shoots for the resources taken up by the host roots, but are relatively weak competitors for light (Matthies, 1995a; Tešitel *et al.*, 2015). These parasites have much higher transpiration rates than their hosts, which they keep up even at night (Press *et al.*, 1988). Negative effects of parasitism on host photosynthesis have also been found and may contribute to reduced productivity (Cameron *et al.*, 2008; Watling and Press, 2001). Because of their negative effects on biomass production, it has been suggested to use root hemiparasites as a conservation management tool to reduce the productivity of grassland communities in nature reserves and thus to increase diversity (Westbury *et al.*, 2008) potentially. Contrary to *Osyris* spp., some allied species in Santalaceae produce edible fruits, and others are regarded as a source of timber and perfume (sandalwood/sandalwood oil) (Watson and Dallwitz, 1992). However, different chemicals were isolated and identified from *Osyris alba,* many of which have medicinal uses for different ailments, including pyrrolizidine alkaloid, flavonoids, and some organic acids (Hamdan and Afifi, 2008; Tsukasa *et. al.*, 2008; Al-Jaber *et al.*, 2010). More important is the presence of quinolizidine and pyrrolizidine alkaloids; that enable the parasite to defend itself against herbivores (Woldemichael and Wink, 2002). In addition, the parasite contains scoparine, genistein, sarothamine and pyrrolizidine flavones, osyrisine, catechin, catechin-3-*O*-α-l-rhamnopyranoside, sparteine, *N*-methylcytosine, cytisine, methyl-12-cytisineacetate, hydroxy-*N*-methylcytosine, *N*-acetylcysteine, lupanine, anagyrine, 1-carboxypyrrolizidine-7-olide, senecionine, integerrimine, retrorsine, senecivernine and the new alkaloid janfestine (7R-hydroxychysin A or 1R-carbomethoxy-7R-hydroxypyrrolizidine). α–amylase, one new pyrrolizidine alkaloid, phytoferritin, P and K, protocatechuic, 4-hydroxybenzoic, p-coumaric, and isoferulic acids are also present as well as rutoside, isoquercitroside, and (after acid hydrolysis) quercetol and kaempferol, D-mannitol, hexaacetate, acetylenic acid and enoic acid. The plant also contains carbohydrates, amino acids, dotriacontanoic and 2-octenoic fatty acids, and phenolics. Engeletin, salvigenin, pachypodol, kumatakillin, penduletin, jecidiene, kaempferol-7-methyl ether, b-sitosteryl glucoside, triterpenoids, ursolic acid, oleanolic acid, and p-hydroxybenzoic acid are also present. mannitol is produced in large amounts (produced as the end product of photosynthesis) while sucrose is typically accumulated in the host. Mannitol biosynthesis from CO2 fixation occurred in aerial parts and especially in the young leaves (Simier *et al.*, 1994; Qasem, 2020b).

Different plant parts are medicinally used. The plant is hypoglycaemic. Its extract is mostly used for controlling glucose after sucrose ingestion and also exhibited the highest inhibitory activity on α-amylase *in vitro.* It is of antibacterial, anti-parasitic, antimicrobial and anti-bleeding activities and is used against edemas. Plant methanol and butanol extracts exhibited a significant level of anti-parasitic activity against *Entamoeba histolytica* and *Giardia intestinalis* and against cancer (Hep-2) and non-malignant (Vero) cells. Extracts are useful in controlling postprandial blood glucose, their effects being mainly achieved through the inhibition of carbohydrate-metabolizing enzymes (amylases and glucosidases). Due to the mechanism involved in controlling blood glucose, such plant extracts are also expected to be helpful in the management of obesity (Qasem, 2020b).

In folk medicine, some herbalists sell the plant for use in the treatment of amoebic dysentery. The leaves and seeds are aphrodisiac and are used for alopecia.

Osyris lanceolata essential oils are used in the manufacture of perfumery and medicinal products in some African countries.

CONTROL

Studies on the control of *Osyris* spp. are generally lacking or very limited. Not much work on these parasites either on their biology, physiology, ecological responses, or economic importance. This literature limitation may be partly due to low attention given to these parasites since they are not obviously observed as serious threatening parasites to cultivated crops and their behavior as facultative parasites found growing by themselves like other nonparasitic species. However, glyphosate (0.8%), sprayed directly onto the parasite's foliage, and 15% triclopyr, and 10% glyphosate, wiped onto the parasite's foliage using a hand rope wipe were tested. The best results were obtained with glyphosate and triclopyr applied by the rope wipe method (Joel *et al.*, 1991). Insects such as lady beetles can live on foliage of certain species and fruits may be attacked by fungus. Control methods for different species of *Osyris* using natural enemies are shown in Table 2.1.1.

Table 2.1.1. Natural enemies of some *Osyris* species

Osyris species	Enemy species	Reference
Osyris abyssinica	*Mylothris agathina*	http://www.funet.fi/pub/sci/ /bio/life/warp/food-plants-o.htm
Osyris alba	*Orgyia trigotephras*	
Osyris compressa	*Charaxes pelias*	
Osyris lanceolata	*Charaxes hansahli* *C. pelias*	
Osyris quadripartita	*Charaxes jasius*	

OSYRIS (*OSYRIS* SPP.) IN JORDAN

GENERAL BACKGROUND

In two studies carried out by Qasem (2006 and 2020a) on *Osyris* in Jordan, the author recorded *Osyris alba* on different hosts just in certain locations in the country (Table 2.1.2). The parasite was recorded and illustrated growing with 23 host species of gymnosperms and angiosperms belonging to 14 plant families of the most parasitized were Leguminosae (4 species), Rosaceae (4 species) and Rhamnaceae (2 species). However, a connection with 18 host plant species was checked and confirmed, while only five were left without being insured because of the difficulty in digging the rocky soil or following the extension of host root systems. These are *Casuarina equisetifolia*, *Cistus creticus*, *Crataegus azarolus*, *Ficus carica*, and *Ziziphus lotus*. The infection of these latter species was judged from the thick vegetative growth of the parasite found close to the trunk area or underneath these species, and also from the association of some of these plants with the parasite in more than one location.

All attacked species are shrubs or fruit and forest trees of economic and/or ecological importance. The study is the first report on *O. alba* as a parasitic species in the Kingdom of Jordan. It provides some information on the potential threat that this parasite might exert on other plant species in certain parts of the country, while the attack of different cultivated fruit species by the parasite is a new development in cultivated fields. It seems that the parasite has a wide host range of wild and cultivated species that differ in growth habits and in habitat (Table 2.1.2). It was found on *Acacia cyanophylla* Lindl., *Prunus domestica* L., and underneath *Ziziphus lotus* L. in the Irano-Turanean region at ≈ 150 m a.s.l., as well as in the Mediterranean region (*e.g.* on *Cupressus sempervirens* L., *Pinus halepensis* Mill., *Sarcopoterium spinosum* (L.), *Olea europaea* L., and *Coridothymus capitatus* (L.) Hoffmanns. et Link) at 800–1100 m a.s.l. in Zay, Amman, and Na'ur. It occurs on sandy-loam and stony soils in both biogeographic regions. These high elevations are characterized by cool weather conditions during the winter, frequently receive some snow, and are exposed to frost for a few days in the winter season. The parasite, however, was not detected in any of the surveyed stations in the Jordan Valley (Sudanian region). These findings appear in conflict with reports that negate the presence of *O. alba* in cold regions and describe it as a frost-sensitive species (Watson and Dallwitz 1992). The diverse host species of *O. alba* might reflect high adaptation of the parasite to different hosts. However, the severity of infection was varied for different species, indicating a possible host preference. Location- and species-mediated differences in the intensity of infestation and attack by the parasite prevailed. These differences were exemplified by *Amygdalus communis*, *C. sempervirens*, *Olea europaea*, *Vitis vinifera*, *Quercus coccifera*, *Rhamnus palaestina*, and *Sarcopoterium spinosum*, which were heavily infected, severely damaged, and attacked in different locations of the surveyed area, while *Pistacia palaestina*, *Rhus coriaria*, *Acacia cyanophylla*, and *Z. lotus* were considerably less parasitized. The intensity of infection (vegetative mass of parasite shoots on or underneath the host plant) was different in the various host species. The parasite was formed thick and has massive vegetation, with long branches (2–4 m), and heavily fruiting on *A. communis*, *O. europaea*, *C. sempervirens*, and *V. vinifera*. On *P. halepensis*, it developed a light infection, with the parasite displaying only a few shoots and clearly short (25–40 cm) branches. Many other plant species, including the trees *Ceratonia siliqua* L., *Arbutus andrachne* L., and *Eucalyptus camaldulensis* Dehnh., were not attacked. Certain species, such as *A. communis*, *C. sempervirens*, and *S. spinosum*, were widely attacked in different locations. The leaves of parasitized *C. sempervirens* plants were yellow, and many of the branches died out. Other host species, including *A. cyanophylla*, *C. equisetifolia*, *F. carica*, *O. europaea*, *P. domestica*, *P. palaestina*, *C. capitatus*, and *S. spinosum*, are the first time reported as hosts for this parasite. The parasite was destructive to *A. communis*, *P. domestica*, *O.*

europaea, and *Calicotome villosa* and seems to have killed these hosts at certain sites. However, the death of the host plants did not appear to prevent the parasite from completing its life cycle successfully. The parasite is able to photosynthesize, and was observed growing normally in arable land in different places of the surveyed locations in the absence of any surrounding vegetation. Local farmers were not familiar with the problem and the damage it causes to crops and trees. Illustrations showing the parasite on different host plant species in Jordan are presented in plates (1-26).

Table 2.1.2. Host species and their families, status and intensity of infection of the *Osyris alba* / Santalaceae found in Jordan.

Scientific name	Family	Growth Form	Status	Intensity of infection[1]	Severity of infection[2]	Biogeographical region
Forest trees						
Acacia cyanophylla Lindl.	Fabaceae	ST	C	M	3	Irano-Turanean
Casuarina equisetifolia L.	Casuarinaceae	T	C	M	1	Mediterranean
Crataegus azarolus L.**	Rosaceae	ST	W	M	2	Mediterranean
Cupressus sempervirens L. *var. horizontalis* (Miller) Gordon	Cupressaceae	T	C	H	4	Mediterranean
Cupressus sempervirens L. *var. pyramidalis* Nyman	Cupressaceae	T	C	H	4	Mediterranean
Pinus halepensis Mill.	Pinaceae	T	C	L	1	Mediterranean
Pistacia palaestina Boiss	Anacardiaceae	ST	W	M	1	Mediterranean
Quercus coccifera L.	Anacardiaceae	T	W	H	2	Mediterranean
Retama raetam (Forskal) Webb and Berth	Fabaceae	S	C	M	1	Irano-Turanean, Mediterranean
Rhamnus palaestina Boiss	Rhamnaceae	S	W	H	4	Mediterranean
Ziziphus lotus L.**	Rhamnaceae	S	W	M	1	Irano-Turanean
Forest shrubs						
Anagyris foetida L.	Fabaceae	S	W	M	2	Mediterranean
Calicotome villosa (Poir) Link.	Fabaceae	S	W	H	4	Mediterranean
Cistus creticus L.**	Cistaceae	S	W	L	1	Mediterranean
Clematis cirrhosa L.	Ranunculaceae	S	W	M	2	Mediterranean
Rhus coriaria L.	Anacardiaceae	S	W	M	1	Mediterranean
Sarcopoterium spinosum (L.) Spach	Rosaceae	S	W	H	5	Mediterranean
Coridothymus capitatus (L.) Hoffmanns. and Link	Lamiaceae	S	W	M	3	Mediterranean

Euphorbia hierosolymitana Boiss	Euphorbiaceae	S	W	L	3	Mediterranean
Fruit trees						
Amygdalus communis L.	Rosaceae	T	C and W	H	4	Mediterranean
Ficus carica L.**	Moraceae	T	C	M	2	Mediterranean
Olea europaea L.	Oleaceae	T	C	H	4	Mediterranean
Prunus domestica L.	Rosaceae	T	C	M	3	Irano-Turanean
Vitis vinifera L.	Vitaceae	S	C	H	4	Mediterranean

**: Connection with the parasite was not checked
Form: S = shrub, ST = small tree, T = tree
Status: C = cultivated, W = wild grown
[1] Intensity of infection (relates to the extent of parasite development): Arbitrary value rating of low (L), moderate (M) or high (H)
[2] Severity of infection: 1 denotes that the host was least affected, and 5 denotes that the damage to host plant was severe.
Source: Qasem, 2006, 2020a

SOME ILLUSTRATIONS ON *OSYRIS ALBA* AND ITS HOSTS IN JORDAN

Plate 1. *Osyris alba* at flowering stage

Plate 2. ***Osyris alba*** at fruiting stage on *Cupressus sempervirns*

OSYRIS ALBA ON WILD SHRUBS

Plate 3. *Osyris alba* on wild almond

Plate 4. *Osyris alba* on *Anagyris foetida*

Plate 5. *Osyris alba* on *Calicotome villosa*

Plate 6. *Osyris alba* killing *Calicotome villosa*

Plate 7. *Osyris alba* spread on *Sarcopoterium spinosum*

Plate 8. *Osyris alba* on *Coridothymus capitatus*

Plate 9. *Osyris alba* at flowering stage on *Coridothymus capitatus*

Plate 10. *Osyris alba* on *Euphorbia hierosolymitana*

Plate 11. *Osyris alba* on *Clematis cirrhosa*

OSYRIS ALBA ON FRUIT AND FOREST TREES

Plate 12. *Osyris alba* on cultivated almond

Plate 13. *Osyris alba* on grapes (*Vitis vinifera*)

Plate 14. *Osyris alba* on olive (*Olea europaea*)

Plate 15. *Osyris alba* on fig (*Ficus carica*)

Plate 16. *Osyris alba* on plum (*Prunus domestica*)

Plate 17. *Osyris alba* on oak (*Quercus coccefera*)

Plate 18. *Osyris alba* on *Rhamnus palaestina*

Plate 19. *Osyris alba* on *Pistacia palaestina*

Plate 20. *Osyris alba* killed *Pistacia palaestina*

Plate 21. *Osyris alba* on *Pinus halepensis*

Plate 22. *Osyris alba* on *Cupressus sempervirens var. horizontalis*

Plate 23. *Osyris alba* on *Cupressus sempervirens var. pyramidalis*

Plate 24. *Osyris alba* on *Retama raetam*

Plate 25. *Osyris alba* on *Acacia cyanophylla*

Plate 26. *Osyris alba* killed *Acacia cyanophylla*

REFERENCES

Al-Jaber, H.I., Mosleh, I.M., Mallouh, A., Abu Salim, O.M., Abu Zarga, M.H. Chemical constituents of *Osyris alba* and their antiparasitic activities.*J Asian Nat Prod Res.,* 2010, 12(9), 814-20. doi: 10.1080/10286020.2010.502892.

Aronne, G., Wilcock, C., Pizzolongo, P. Pollination biology and sexual differentiation of *Osyris alba* (Santalaceae) in the Mediterranean region. *Plant Syst. Evol.,* 1993,188 (2), 1-16.

Benharrat, H., Renaudin, S., Rey, L., Thalouarn, P. Sur la différenciation des trachéides dans les haustoriums d' *Osyris alba* parasitant *Hedera helix. Can J Bot,* 1987, 65 (8), 1746-1755. Available at: https://doi.org/10.1139/b87-239.

Cameron, D.D., Geniez, J.M., Seel, W.E., Irving, L.J., Suppression of host photosynthesis by the parasitic plant *Rhinanthus minor. Ann. Bot.,* 2008,101, 573-578.

Hamdan, I.I., Afifi, F.U. Screening of Jordanian flora for alpha-amylase inhibitory activity. *Pharm Biol,* 2008, 46, 10-11, 746-750. Available at: http://dx.doi.org/10.1080/13880200802316053.

Herrera, C. M. Predispersal reproductive biology of female *Osyris quadripartita* (Santalaceae), a hemiparasitic dioecious shrub of Mediterranean scrublands. *Bot. J. Linn. Soc.*, 1985, 90, 113-127. Available at: http:www.jstor.org/.

Herrera, C. M. The annual cycle of *Osyris quadripartita*, a hemiparasitic dioecious shrub of Mediterranean scrublands. *J. Ecology*, 1984, 72, 1065-1078. Available at: http:www.jstor.org/.

Herrera, C.M. Plant size, spacing patterns and host plant selection in *Osyris quadripartita*, a hemiparasitic dioecious shrub. *J. Ecol.*, 1988b,76 (4), 995-1006. Available at: http://www.ebd.csic.es/info/verresum? P. 1364.

Herrera, C.M. The fruiting ecology of *Osyris quadripartita*: individual variation and evolutionary potential. *Ecology*, 1988a, 69 (1), 233-249. Available at: http://www.ebd.csic.es/info/verresum.

Joel, D. M., Kleifeld, Y., Bucsbaum, H. *Osyris alba* causing damage in orchards. In: Ransom, J.K., Musselman, L.J., Worsham, A.D., Parker, C., (eds.), *Proceedings of the 5th International Symposium of Parasitic Weeds.* Nairobi, Kenya, 24-30 June 1991, pp. 378-381.

Matthies, D. Parasitic and competitive interactions between the hemiparasites *Rhinanthus serotinus* and *Odontites rubra* and their host *Medicago sativa. J. Ecol.*, 1995a, 83, 245-251.

Mwang'ingo, P.L., Telklehaimanot, Z., Lulandula, L.L., Mwihomeke, S.T. Host plants of *Osyris lanceolata* (African Sandalwood) and their influence on its early growth performance in Tanzania. *South. Afr. For. J.,* 2005, 203, 55-65.

Nickrent, D.L. Parasitic Plants of the World. Chapter 2, pp. 7-27. In: López-Sáez, J. A., Catalán. P., Sáez, L., (eds.), *Parasitic Plants of the Iberian Peninsula and Balearic Islands.*, Mundi-Prensa, Madrid. 2002.

Nickrent, D.L., Musselman, L.J. Introduction to Parasitic Flowering Plants. *The Plant Health Instr.,* 2004. DOI: 10.1094/PHI-I-2004-0330-01.

Planchon, J.E. 1858. On the parasitism of *Osyris alba. Ann. mag. nat. hist. series* 3, 2: 254-255. 2265139, 10.1080/00222935808697021

Press, M.C., Graves, J.D., Stewart, G.R. Transpiration and carbon acquisition in root hemiparasitic angiosperms. ***J.*** *Exp. Bot.,* 1988,39,1009–1014.

Qasem, J.R. Host range of the parasitic weed *Osyris alba* L. in Jordan. *Weed Biol. Manag.*, 2006, 6 (2), 74-78.

Qasem, J.R. Osyris (*Osyris alba* L) Updates, the illustrated threatening parasite to fruit and forestry trees in Jordan. *Pak.J.Botany*, 2020a, 52(1), 251-256. DOI: 10.30848/PJB2020-1(21).

Qasem, J.R. *The Coloured Atlas of Medicinal and Aromatic Plants of Jordan and Their Uses* (volume 3). Cambridge Scholar Publishing, Newcastle, UK. 2020b.

Simier, P., Renaudin, S., Fer, A. Characteristics of the mannitol pathway in a root hemiparasitic species, *Thesium humile* Vahl. (Santalaceae). *J. Plant Physiol.*, 1994, 143(1), 33-38.

Southern Illinois University Carbondale (SIUC). Santalaceae (including Eremolepidaceae), Family description. 2005. Available at: http://www.parasiticplants.siu.edu/Santalaceae/index.html. Accessed 16 March 2006.

Těšitel, J., Těšitelová, T., Fisher, J. P., Lepš, J., Cameron, D. D. Integrating ecology and physiology of root-hemiparasitic interaction: interactive effects of abiotic resources shape the interplay between parasitism and autotrophy. *New Phytol.*, 2015, 205, 350-360.

Tsukasa, I., López-Sáez, J A., Junichi, K. Flavonoids from *Osyris alba*. *Biochem. Syst. Ecol.* 2008, 36(2),146-147. DOI: 10.1016/j.bse.2007.06.

Watling, J.R., Press, M.C. Impacts of infection by parasitic angiosperms on host photosynthesis. *Plant Biol.,* 2001, 3, 244-250.

Watson, L., Dallwitz, M. J. *The Families of Flowering Plants: Descriptions, Illustrations, Identification, and Information Retrieval.* 1992 Version: 14th December 2000. Available at: http://biodiversity.uno.edu/delta/.

Westbury, D.B., Dunnett, N.P. The promotion of grassland forb abundance: A chemical or biological solution? *Basic Appl. Ecol.*, 2008, 9, 653-662.

Woldemichael, G.M., Wink, M. Concomitant occurrence of pyrrolizidine and quinolizidine alkaloids in the hemiparasite *Osyris alba* L. (Santalaceae). *Biochem. Syst. Ecol.,* 2002, 30 (2), 139-149.

Websites

http://www.funet.fi/pub/sci//bio/life/warp/food-plants-o.htm

CHAPTER 4

Thesium (*Thesium* spp.)

Jamal Ragheb Said Qasem

INTRODUCTION

Thesium spp. are annuals or perennials that belong to the Santalaceae family. They are root-hemiparasitic weeds, have chlorophyll, can do photosynthesis and produce carbohydrates but can also parasitize hosts of different species, mainly monocots.

Dwarf thesium (*Thesium humile* Vahl) (Synonyms *Thesium dollineri*) is an annual reproduces by seeds. It is 10-30 cm in height, a facultative hemiparasite contains chlorophyll pigments and can do photosynthesis but can parasitize different plant species. It is an erect plant or ascending and furrowed angled. Leaves are dense, narrowly linear, 1-nerved, and rough, especially at the margin; bracts very long, leaf-like bracteoles much shorter than bracts but longer than flowers. Flower segments pale green, fruit is a nutlet, 6 mm, ovoid to subglobular, netveined between nerves. The plant flowers from February to May. It is distributed in cultivated fields, parasitizing barley and onion in certain locations in Jordan.

GERMINATION AND DEVELOPMENT

Thesium humile seed is of a very hard coat (Fig. 2.2.1) that requires scarification or invasion by fungi to enhance its germination by partial destruction of the hard coat.

Fruits

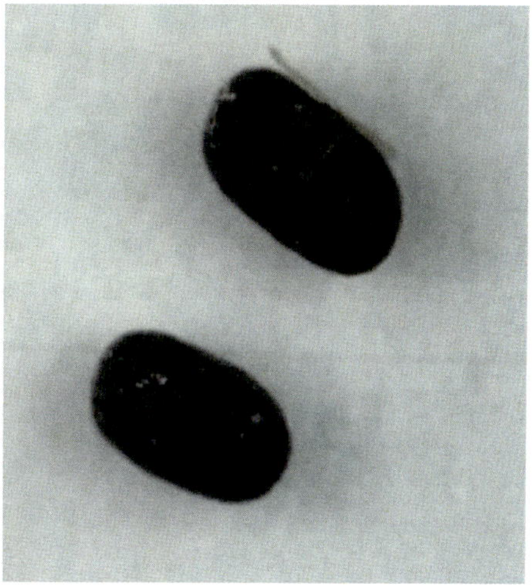

Seeds

Fig. 2.2.1. *Thesium humile* fruit capsules and seeds

Fruit coat is adapted to retain water under very dry conditions. *Thesium* as a genus may have developed a broad host range and yet has strains that have developed extremely narrow parameters for host selection and may be better recognizing host root exudates (Holzner and Numata, 1982).

Thesium humile Vahl (Santalaceae) is a herbaceous, obligate root hemiparasitic angiosperm, can live autotrophic for several weeks before attachment to the host. The parasite showed a decreasing growth rate and symptoms of senescence of the seedling at about 6 weeks after germination. Before attachment, *Thesium* is able to take up water and inorganic ions from the soil, but its mineral content is characterized by very high Na and low P levels. During the same period, rates of net photosynthesis and dark respiration, as well as RuBPc and PEPc activities, decrease continuously (Fer *et al.*, 1994). After attachment to the host, *Thesium* recovers a more physiological balance by deriving inorganic ions from the host xylem sap. Increasing growth rate, photosynthetic carbon fixation, carboxylase capacities and expression of a photosynthetic gene like rbcL are observed and become fairly similar to those found in non-parasitic C-3 plants. A strong K accumulation in leaves also occurred and that might be an explanation for the observed anomalous stomatal behavior. Before attachment, the poorly developed root system of *Thesium* seems unable to selectively absorb inorganic ions from the soil and to maintain a physiological mineral balance. Consequential to the very high Na and the low P contents in the seedling of the parasite, metabolic perturbations also affecting gene expression might result. (Fer *et al.*, 1994).

CONTACT AND ATTACHMENT

Thesium humile var. maritima is a hemiparasite angiosperm that attaches to a wide range of host plants, Development of the haustorium on *Hordeum vulgare* is described and illustrated as is the mature haustorium on *Plantago albicans*. Despite minor differences, the *Thesium* haustorium shows all the characteristics typical of Santalaceae. In *Thesium*, autoparasitism is common, and self-haustoria, connecting either with the roots of different *Thesium* plants or with the same plant were also recorded (Abou-Raya *et al.*, 1977). However, the existence of haustoria attached to the roots of a plant does not mean that these haustoria are functional because the host plants may block them (Cameron and Seel, 2007; Govier, 1966; Rümer *et al.*, 2007).

BIOLOGY, ECOLOGY AND PHYSIOLOGY

Thesium humile stems are 5-20 cm, ascending to erect, usually branched from the base with long, simple branches. Leaves are linear or narrowly linear, rather fleshy, denticulate, and 1-veined. Inflorescence is often subspicate, sometimes paniculate. Bract 2-5 times as long as flower or fruit; bracteoles usually equaling flower or fruit. Perianth infundibuliform, 5-lobed. Nut subglobose, reticulated veined, 5-6 times as long as the persistent perianth.*Thesium* spp. possess chlorophyll but obtain some of their nutritional needs from other plants. Species in the genus are often unattractive, having a yellowish colour and lacking leaves. The flowers are frequently small and creamy-white, and most species are probably pollinated by small bees and flies (Hendrych, 1972).The weed parasitizes cereals and mainly wheat and barley.*Thesium* physiology has been studied in an agronomic context. In particular, mannitol metabolism has been thoroughly studied for the purpose of controlling its weedy habit (Fer *et al.*, 1993; Simier *et al.*, 1993, 1994, 1998; Williamson *et al.*, 2002). Very little is known on parasite-host range and selectivity in wild populations. *Thesium chinense* is thought to parasitize species of Poaceae (Numata and Yoshizawa, 1975), but the information is based on a limited number of observations and circumstantial evidence that these plants usually occur near plants from the family Poaceae. The situation is more or less the same in other *Thesium* species, but the genus as a whole seems capable of using a wide range of angiosperms, including *Themeda* and *Poa* (Poaceae; Scarlett *et al.*, 2003), *Galium* (Rubiaceae; Renaudin *et al.*, 1981), *Hordeum vulgare* and *Allium cepa* (Abu-Irmaileh, 1980) and grape (Dasgupta, 1988). Therefore, it is possible that each *Thesium* species parasitizes several host species within a population while having some preference for particular groups of hosts (*e.g.* Poaceae).

DISTRIBUTION AND HOST RANGE

Thesium (The Sandalwood family) is a genus of herbaceous and woody root hemiparasites widely distributed in the temperate and tropical regions of the old world (Pilger, 1935). The genus consists of almost 200 species, chiefly native to Africa and the Mediterranean region (Engler,1964; Cronquist, 1981). Mabberley (2008) mentioned 300 species with approximately 150 are native to southern Africa,with 81 species native in the Cape or Cape Floristic Region (Manning and Goldblatt, 2000), including 35 endemics. A further 60 species occur in tropical and northern Africa, and the rest of the genus is primarily distributed in Europe and Asia. Only three species occur in South America, with two species centered in Brazil and a third native to Venezuela. Some species are well-known agricultural weeds in Africa, Australia, Czechoslovakia, France, Egypt, Greece, Maltese Islands, Jordan, Portugal, Spain, Turkey and USA, and with 18 species occur in Europe, the Balearic Islands, Corse, and Crete, Italy.

Species of *Theism* appear to parasitize a wide range of angiosperms as hosts, including *Felicia muriatic, Felicia filifolia, Chrysocoma tenuifolia, Pteronia sordida, Melianthus comosus* and *Lycium* spp. (Steyn, 1949; Varhmeijer, 1981). *Themeda* Forssk. and *Poa* L., (Scarlett *et al.*, 2003); *Galium* L., (Renaudin *et al.*, 1981); *Hordeum* L. and *Allium* L. (Abu-Irmaileh, 1980); and *Vitis* L., (Dasgupta, 1988). Therefore, it is possible that each *Thesium* species parasitizes several host species within a population while having some preference for particular groups of hosts (*e.g.* Poaceae). It has been suggested that *Thesium* spp. do not specifically select hosts, but rather occur in microhabitats with specific conditions (Dostálek and Münzbergová, 2010). It is found in cultivated fields, parasitizing barley in different parts of the country and attacking onion in certain places in the Jordan valley. The parasite weakening host plants and mainly barley. *Thesium humile* was found on roots of *Triticum vulgare* (Fer *et al.*, 1994), and *Thesium humile var. maritima* parasitizing *Artemisia campestris, Bromus rigidus, Lolium perenne, Setaria glauca, Allium cepa* and *Hordeum vulgare* (Abou-Raya *et al.*, 1977).

Thesium chinense is another perennial parasitizing *Gnaphalium affine, Chrysanthemum indicum, Artemisia japonica, Triticum aestivum, Imperata cylindrica, Prunella vulgaris, Eremochloa ophiuroides,* and *Cynodon dactylon* (Guo and Luo, 2010).

ECONOMIC IMPORTANCE

Thesium humile is a small, herbaceous hemiparasitic angiosperm, the endemic herb in North Africa and probably in other Mediterranean countries (Simier *et al.*, 1994). It parasitizes several host species of Poaceae, Fabaceae, Liliaceae, Rubiaceae and others (Chabrolin, 1934; Qasem, 2010), showing some sort of host specificity between these. In addition to causing important yield losses in cereal crops (El Hamri, 1988), it is also responsible on animal poisoning, especially among small ruminants (Bellakhdar, 2008). It was reported to cause very acute poisonings since the death of intoxicated animals occurs shortly after feeding (Belakhdar *et al.*, 2014).

Thesium humile Vahl. causes damage in cereal crops in the Mediterranean countries to associate with wheat, barley, onion, and sugarcane. The K, Ca, Na and Mg contents of *Thesium humile* were found to be much greater than those of its host (*Triticum vulgare* and *T. aestivum*), although ratios of the concentration of different ions were very similar in both plants; it is suggested that the root hemiparasite obtains its mineral nutrients mainly from the host xylem (Fer *et al.*, 1992). *Thesium humile* contains large amounts of mannitol (produced as the end product of photosynthesis) while sucrose is typically accumulated in the host (Simier *et al.*, 1994). Seed oils, alkaloids including 1-hydroxymethyl pyrrolizidine, fatty acid esters, phenols, alkanes and alkenes. The plant has antimicrobial, antifungal, antioxidant, anti-cancer, antiphlogistic and hypo-cholesterolemic, and anti-inflammatory properties. The plant is considered toxic to small ruminants (Qasem, 2020). It has been reported as caused cattle poisoning in Morocco, where this plant is naturally abundant and certain species of the genus have been suspected of causing death to livestock in South Africa.

CONTROL

Literature on the control of *Thesium* in general and of *Thesium humile* is scanty. However, *T. humile* may be easily controlled by hand weeding, which must be practiced early after parasite emergence and before or after attachment to the host plant but always before flowering and seed production. In addition, herbicides of the phenoxy group are usually used in cereal crops as broadleaf killers and other contact herbicides selectively used in onions as pre-plant soil-applied including pendimethalin or post-plating foliage applied such as oxyfluorfen and oxadiazon herbicides may be tried. In addition, *Thesium linophyllon* has also been reported to be attacked by *Cuscuta epithymum* that indicates a kind of epiparasitism is also exists between both parasites (Piwowarczyk *et al.*, 2017). The epiparasite may harm the host-parasite relatively more than the same parasite would harm a nonparasitic host (Heide-Jorgensen, 2008).

THESIUM (*THESIUM* SPP.) IN JORDAN

GENERAL BACKGROUND

Two species of *Thesium* have been reported to occur in Jordan. These are *Thesium bergeri* Zucc and *Thesium humile* Vahl. While no records on the parasitism or hosts of the first species on economic crops or wild-grown species are available, two host species were documented by different authors for the second species. These were *Hordeum vulgare* and *Allium cepa* in King Talal Dam area and Al-Karamah, Jordan valley, respectively (Abu-Irmaileh, 1980; Qasem, 2010). Some illustrations on *Thesium* spp. in Jordan and their hosts are shown in Plates (1-4).

SOME ILLUSTRATIONS ON *THESIUM* SPP. AND THEIR HOSTS IN JORDAN

Plate 1. *Thesium humile* on onion (*Allium cepa*)

Plate 2. *Thesium humile* on barley (*Hordeum vulgare*)

Plate 3. *Thesium humile* on barley (*Hordeum vulgare*)

Plate 4. *Thesium bergeri* Zucc.

REFERENCES

Abou-Raya, M. A., El-Sharkawy, M. A., Garaboli, F. R. Observations on the parasitic behavior of *Thesium* Vahl. (Santalaceae). (a) hosts and mode of attachment of haustorial structure. *Comptes Rendus du Ve Colloque International sur l'Ecologie et la Biologie des Mauvaises Herbes, Dijon*, 1976, 1977. 367-376.

Abu-Irmaileh, B.E. *Thesium humile* on onion. *Haustorium,* 1980, 6, 2.

Belakhdar, G., Benjouad, A., Kessabi, M., Abdennebi, E.H. Identification of the pyrrolizidine alkaloid 1-hydroxy methyl pyrrolizidine from *Thesium humile* Vahl. *J. Mater. Environ. Sci.,* 2014, 5 (3), 811-814.

Bellakhdar, J. Hommes et plantes au Maghreb-Eléments pour une méthode en ethnobotanique, (Metz): Plurimondes, cop. 2008.

Cameron, D.D., Seel, W.E. Functional anatomy of haustoria formed by *Rhinanthus minor*: linking evidence from histology and isotope tracing. *New Phytol.* , 2007, 174, 412-419.

Chabrolin, C. *Annales du Service Botanique de Tunisie,* 1934,11, 65.

Cronquist, A. *An Integrated System of Classification of Flowering Plants.* Columbia Univ., New York. 1981.

Dasgupta, M.K. *Principles of Plant Pathology*. New Delhi: Allied Publisher. 1988.

Dostálek, T., Münzbergová, Z. Habitat requirements and host selectivity of *Thesium* species (Santalaceae). *Bot. J. Linn.Soc.*, 2010, 164, 394–408.

El Hamri, D. *Mémoire d'Ingénieur d'Application*, IAV Hassan II, Rabat, 1988, 72 p.

Engler, A. *Syllabus der pflanzenfamilien*, 1964, 12 ed. Vol. 2.-Gebr. Borntraeger, Berlin.

Fer, A., Painset, J. C., Rey, L. Characteristics of mineral nutrition in some types of parasitic plants (*Cuscuta, Orobanche, Thesium*) and their development on crops. *Bulletin de la Societe Botanique de France, Actualites Botaniques,* 1992, 139(1), 111-121.

Fer, A., Russo, N., Simier, P., Arnaud, M.C., Thalouarn, P. Physiological changes in a root hemiparasitic angiosperm, *Thesium humile* (Santalaceae), before and after Attachment to the Host Plant (*Triticum vulgare*).*J. Plant Physiol.,*1994. 143 (6), 704-710. Available at: https://doi.org/10.1016/S0176-1617(11)81161-2.

Fer, A., Simier, P., Arnaud, M., Rey, L., Renaudin, S. Carbon acquisition and metabolism in a root hemiparasitic angiosperm, *Thesium humile* (Santalaceae) growing on wheat (*Triticum vulgare*). *Funct. Plant Biol.,* 1993,20, 15–24.

Govier, R.N. *The Interrelationships of the Hemiparasites and Their Hosts, with Special Reference to* Odontites verna (Bell. Dum). Ph.D. Thesis, University of Wales. 1966.

Guo, Q.S., Luo, F.L. Comparative studies on the growth, chlorophyll, amino acids and minerals of *Thesium chinense* (Santalaceae) in association with different hosts. *Nord. J. Bot.*, 2010, 28, 632-640.

Heide-Jorgensen, H. S. *Parasitic Flowering Plants*. Brill, Leiden, The Netherlands, 2008. ISBN: 978-90-04-16750-6. Available at: https://doi.org/10.1163/ej.9789004167506.i-438

Hendrych, R. *The Natural History and Systematics of Genus* Thesium. Charls University, Department of Botany, 1972. or at: *Acta Univ, Carolinae* (*Biol.*). *Tchecosl; DA,* 1970, 4, 239-358.

Holzner, W., Numata, M. (eds.). *Biology and Ecology of Weeds*. Dr. W. Junk Publishers. The Hague-Boston-London. 1982.

Mabberley, D. *Mabberley's Plant-Book. A Portable Dictionary of Plants, Their Classifications, and Uses*, 3rd ed. Cambridge: Cambridge University Press. 2008.

Manning, J.C., Goldblatt, P. *Cape Plants. A Conspectus of the Cape Flora of South Africa. Strelitzia 9*. Pretoria: National Botanical Institute and Missouri Botanical Garden Press. 2000.

Numata, S., Yoshizawa, N. *Weed Flora of Japan*. Tokyo: Zenkoku Noson Kyoiku Kyokai; 1975.

Pilger, R. Santalaceae. In: Engler, A., Harms, H., (eds.), *Dienaturlichen Pflanzenfamilien*, 16. Leipzig: Duncker & Humblot, 1935, 52–91.

Piwowarczyk, R., Guzikowski, S., Goralski, G., Denysenko-Bennett, M., Kwolek, D., Joachimiak, A.J. First report of dodder (*Cuscuta epithymum*) parasitizing hemiparasitic species of Santalaceae (*Thesium*) and Orobanchaceae (*Euphrasia, Melampyrum, Odontites, Orthantha,* and *Rhinanthus*) in Poland. *Plant Dis.,* 2017, 102 (2), 456. Available at: https://doi.org/10.1094/PDIS-03-17-0389-PDN.

Qasem, J. R. *The Coloured Atlas of Medicinal and Aromatic Plants of Jordan and Their Uses*. Volume Three. 2020. Cambridge Scholars Publishing. Newcastle, UK. 491 PP. ISBN 978-1-5275-4561X.

Qasem, J.R. Parasitic flowering plants of cultivated plants in Jordan, the present status and possible management. *Pak. J. Weed Sci. Res.*, 2010,16(2), 227-239.

Renaudin, S., Cheguillaume, N., Gallant, D.J. Distribution and role of mineral compounds in the haustorium of a parasite of *Galium arenarium, Thesium humifusum*, before flowering. *Can. J. Bot.* 1981,59, 1998–2002.

Rümer, S., Cameron, D.D., Wacker, R., Hartung, W., Jiang, F. An anatomical study of the haustoria of *Rhinanthus minor* attached to roots of different hosts. *Flora*, 2007, 202, 194-200.

Scarlett, N, Bramwell, M., Earl, G. Austral toad flax. *Flora and Fauna Guarantee Action Statement,* 2003,56, 1–5.

Simier, P., Fer, A., Renaudin, S. Identification of the main osmotically active solutes in the unstressed and water-stressed root-hemiparasitic angiosperm *Thesium humile* and its host *Triticum vulgare*. *Funct. Plant Biology*, 1993, 20, 223–230.

Simier, P., Renaudin, S., Fer, A. Characteristics of the mannitol pathway in a root hemiparasitic species, *Thesium humile* Vahl. (Santalaceae). *J. Plant Physiol.*, 1994, 143 (1), 33-38.

Simier, P., Robert, S., Fer, A. Mannitol metabolism in darkness in the leaves of the hemiparasitic angiosperm. *Thesium humile*. *Plant Physiol. Bioch.*, 1998,36, 237–245.

Steyn, D. G. Vergiftiging van mens en dier. Pretoria: Van Schaik. 1949.

Vahrmeuer, J. *Poisonous Plants in Southern Africa*. Cape Town: Tafelberg. Printed. 1981.

Williamson, J.D., Jennings, D.B., Guo, W.W., Pharr, D.M., Ehrenshaft, M. Sugar alcohols, salt stress, and fungal resistance: polyols-multifunctional plant protection? *J. Am. Soc. Hortic. Sci.*, 2002,127,467–473.

SECTION III

THE FAMILY: CYNOMORIACEAE (BALANOPHORACEAE)

Parasitic Weeds of Jordan, 2022, *Part 1*, 317-335

CHAPTER 5

Cynomorium (*Cynomorium coccineum*)

Jamal Ragheb Said Qasem

INTRODUCTION

Parasites of the Cynomoriacea family represent a group of uncertain placement and taxonomic debates on whether the *Cynomorium* genera belong to the family Balanophoraceae (Hooker, 1856; Cronquist, 1981) or the Cynomoriaceae (Takhtajan, 1997). Balanophoraceae includes 18 genus and 44 species in the neo- and paleotropics, whereas Cynomoriaceae includes the *Cynomorium* genus that has two species, *C. songaricum* in eastern Asia and *C. coccineum* in the Mediterranean. The presence of bisexual flowers, an unusual bimodal karyotype (Pazy *et al.*, 1996) and features of the stamens, ovules, and embryo sac (Teryokhin *et al.*, 1975) in *Cynomorium* all support its segregation from Balanophoraceae. Whether considered one or two families, past and present classifications accept relationships between *Cynomorium* and Balanophoraceae until the DNA studies suggest that the genus *Cynomorium* is not a member of this family as was first thought, but more likely belongs to the Saxifragales, possibly close to Crassulaceae (stonecrop family) (Nickrent *et al.,* 2005; Angiosperm Phylogeny Group, 2016). These authors have indicated that *Cynomorium* and Balanophoraceae are not closely related as stated in the past and present classifications. The issue is complicated by the massive horizontal gene transfer between *Cynomorium* and its different hosts (Cusimano and Renner, 2019). The genus *Cynomorium* has been regarded by some workers as a single species of 2 subspecies, *coccineum* occurring in the Mediterranean east to Iran, and *songaricum* (Rupr.) J. Léonard in Central Asia, where is often recognized as a separate species. subsp. *Cynomorium songaricum* (Synonym subsp. *Cynomorium purpureum* Rupr.) and *Cynomorium purpureum* Rupr. (*Cynomorium coccineum* L.) which has a further 1 synonyms. *Cynomorium coccineum* subsp. *coccineum* and *Cynomorium coccineum* subsp. *songaricum* (Rupr.) J. Léonard. Although *Cynomorium* was finally regarded to include two species; *Cynomorium songaricum* and *Cynomorium coccineum.* Ubiquitous in the Mediterranean basin and Arabian Peninsula (Dharmananda, 2011), A more recent study regarded *Cynomorium* in the family Cynomoriaceae and the genus consists of one species, *Cynomorium coccineum* (Cusimano and Renner, 2019), although one of its subspecies is sometimes treated as a separate species (Bellot *et al.*, 2016, Christenhusz and Byng, 2016). Its placement in the Saxifragales was in 2016 with the help of nuclear, plastid, and mitochondrial sequences obtained from next-generation sequencing.

Cynomorium plants are rare and poorly understood parasitic flowering plants, geophytes, herbaceous holo-root parasites, fleshy, monoecious, or dioecious often produce swollen tuberous (haustoria) that connect them to the host plants. Their stems are unbranched, thick and simple, bear numerous leaves, terminated, contracted, opposite or alternating, scale-like and unisexual flowers representing the ultimate reduction among angiosperms (Kuijt, 1969). These parasites are devoid chlorophyll and thus are non-photosynthetic plants; inflorescence is pinkish in color, consists of a dense, erect, club-shaped mass, of 15–30 cm long, carries minute scarlet flowers, which may be male, female or hermaphrodite (UBC Botanical Garden, 2011). It is pollinated by flies, attracted to the plant by its sweet, slightly cabbage-like odor. Once pollinated, the spike turns black. (Dharmananda, 2011). The fruit is a small, indehiscent nut (Fig. 3.1.1).
Cynomorium was mentioned as a parasite of salt-tolerant plants of the Cistaceae family or Amaranthaceae in the Mediterranean region; Amaranthaceae, Tamaricaceae in other places and Nitrariaceae in China, especially *Nitraria sibirica* (Luo *et al.*, 1993). Other authorities suggest the host plants are saltbushes *Atriplex* spp. (Dharmananda, 2011).

In the Arab countries, *Cynomorium* is commonly known by different names, including Tarthuth, Halook Malta, Mazroor, Marsoos, Marshoosh, Tarthooth Profit Adam, Abu Shall, Thairoor,

Fig. 3.1.1. Seeds of *Cynomorium coccineum*

Halook al-Naka'a, Al-Naka'a or Zobb el Ard and also has different English (Common) names including the misleading Maltese fungus; Desert thumb, Red thumb, Tarthuth (Bedouin), Suoyang (Chinese) and Scarlet *Cynomorium* or Maltese mushroom in the western countries (Cui *et al.,* 2013; Dharmananda, 2011). The parasite totally lacks chlorophyll. It is a rare or local species, grows in dry, rocky or sandy soils, often in salt marshes or other saline habitats close to the coast or in the desert in Jordan. It has a wide variety of uses in European, Arabian and Chinese herbal medicine (Qasem, 2020).

GERMINATION AND DEVELOPMENT

Generally, studies devoted to *Cynomorium* seed germination and development under natural conditions are scarce or largely undocumented. Seeds of *Cynomorium* species, after being dropped on the soil are usually dispersed with sand and few reach host roots. Since these species are holo-parasites, they are completely dependent on their hosts for growth factors and do not germinate in nature unless their hosts are available, thus require host stimulant/s in the form of hormones like strigolactones to stimulate germination of their seeds (Kruh *et al.*, 2017). *Cynomorium coccineum* seeds show dormancy which is broken by chemical signals of host roots. The tiny small seeds ready to germinate, start germination producing primary roots grow and extend after seeds receive germination stimulants in root exudates of host plants. However, parasite regeneration from seed that is very difficult to germinate due to the secretion of germination inhibitors such as abscisic acid and presence of thick wax impermeable coat or prevents water and air passage essential for seed germination into the embryo (Dong *et al.*, 2010; Chen *et al.*, 2011). It has been reported that five years or more may be required for successful seed germination and development of *C. songaricum*, under natural conditions. *Cynomorium* seed germination and dormancy breaking may require breaking seed cover with ultrasonic treatment, chemical removal of seed cover using sodium hydroxide and potassium permanganate, seed treatment with gibberellins (GA) and naphthylacetic acid (NAA), or with host root extracts, and/or a combination of all above methods (Luo *et al.*, 2013). Endophytic fungi have been mentioned to have a role and exchanged between *C. songaricum* and its host *Nitraria tangutorum* probably through haustorium, the connection of xylem and phloem in the vascular system (Cui *et al.*, 2018). Some of these fungi showed ability in promoting *C. songaricum* seed germination. However, seed germination of *C. songaricum* was induced in the absence of the host (*Nitraria* sp.) using a mixture extracted with organic solvents and deionized water from the paste of host roots, auxins (IAA and NAA), and cytokinins (benzyladenine). Treatment of seeds with the acetone and ether extracts resulted in 54 and 16% seed germination, respectively. The combinations of auxin (NAA at 1.0 ppm) and cytokinin (benzyladenine of 0.1 ppm) resulted in the greatest seed germination (68%) of *C. songaricum*. The optimum conditions for *C. songaricum* seeds germination were 25-30°C, darkness and pH 6-9. A low-temperature treatment resulted in a high germination rate and short germination time (Luo *et al.*, 1993).

CONTACT AND ATTACHMENT

Upon receiving the germination stimulant/s from roots of host plant and seed germination, the germinated parasite seeds invade the host root, the roots of the parasite are formed and then its rhizomes that produce intrusive absorptive organs "Haustoria" that penetrate later into the root tissues of host plant connecting the parasite with the vascular system of its host. The haustoria begin growing, forming tubercles (rhizomes) on host plant roots attaching the parasite and forming something like a bridge through which materials required for parasite growth are directly absorbed from the host tissues. Under suitable temperature and water conditions from March to early April, the tubercle begins to develop quickly into unbranched stem grow and elongate through the soil, and finally appears above the soil surface as a dark-red or brown stem that starts flowering rapidly, giving small minute brown flowers at a high number difficult to see by naked eyes. The emerged stem is a low-growing inflorescence that usually appears late in the spring. After almost one month, the tissue above ground becomes senile and forms numerous mature spores like seeds upon pollination by flies, then drop and disperse in the desert sand. However, the underground rhizome stays connected to the host root remaining dormant for several years until the next development cycle begins, which concurs in autumn and winter. Endophytic fungi occur with *C. songaricum's* in the desert ecosystem are also play a vital role in parasite development and physiology.

BIOLOGY, ECOLOGY AND PHYSIOLOGY

According to Watson and Dallwitz (1992), the Cynomoriaceae (Balanophoraceae) plants are described as red herbs with thick, simple flowering stems, attack shrubs and trees in wetlands (humid tropics and subtropics), including *Hevea*, *Atriplex* and *Coffea*. They are of the very unusual vegetative form (without chlorophyll, unbranched stems with a reduced alternate; membranous leaves (reduced to deltoid scales). *Cynomorium* is a herbaceous, holo-root parasite, lacks chlorophyll, annual or perennial, and has pours or stomata. Plants have roots, succulent, perennials, rhizomatous (the rhizomes like tubers with absorptive organs called haustoria), fleshy fungus-like, dark red holoparasite. Its primary conductive tissues are central; secondary growth is absent, the terminals of conductive tissues walls are simple, the inflorescence is cylindrical, terminal, crimson, clavate (phylloid), composed of numerous suppressed false capitula combs, brown-reddish, or pinkish in color (Terekhin, 1975), the stem is18-30 cm long above the soil surface but up to 80 cm below the soil, wrongly thought as a fungus and usually soft in texture. Stem is covered by scale-like, reddish leaves that do not bear stomata. The red color of the scales and flowers is due to flavonoids (vide infra) (Harborne *et al.*, 1994). These scales become scarce towards the inflorescence, which itself comprises hundreds of reddish staminate, carpellate, and, in a lower number, bisexual flowers. *Cynomorium coccineum* is, therefore, most often

monoecious. Flowers are dense, zygomorphic, compacted, dioeciously (male and female flowers are separated), rarely hermaphrodite, and plants are monoecious. Flowers are generally minute; aggregate in inflorescence looks like a spike of 7-13 cm long or in heads (Fig. 3.1.2a); the male flowers have small pestles. The staminate flower is made of perianth 5-6 parts, fleshy from the internal side, thick and of 4-7 mm in length, forming a whorl or irregularly spiral arranged below a single stamen. The stamen filament nests in a longitudinal groove on the inner side of a semi-cylindrical or wedge-shaped pistillode with a truncated or notched apex. The pollen of *Cynomorium* is tricolporate (Nickrent *et al.*, 2005). The female flower has a single pestle. The ovary has separated carpels, indehiscent. The carpel is with an elongated, grooved style, and its perianth is reduced to 1–8 small, free papillae at the summit of the ovary or along its sides. The flowers are pollinated by insects, mainly ants and flies; the fruit is hard, closed, looks like a small nut and is usually gathered together (Fig.3.1.2b). The seed has an endosperm and a hard seed coat, the embryo is rudimentary in mature seeds. *Cynomorium songaricum* flowering-fruiting season has been described to be from May to August (Wan and Chen, 2000).
The harmful effects of the parasite result from complete direct parasitism on host plant species, exhausting hosts and causes a great reduction in growth and productivity under intense infestation and death of weak hosts under heavy and repeated infestation. Reports on the infection of these parasites to crop plants and fruit trees however are lacking.

(a) (b)

Fig. 3.1.2. *Cynomortium coccineum* flowering (a) and seeding (b) stems

DISTRIBUTION AND HOST RANGE

The genus *Cynomorium* is generally distributed in temperate to sub-tropical regions, mainly in the Mediterranean and West Asia (Watson and Dallwitz, 1992). It is of one genus having 2 species and is generally spread in South Europe, North Africa, Central Asia (Abd El-Rahman *et al.*, 1999; Duke *et al.*, 2008; Dharmananda, 2011) and probably other places in the world. The representative species is *Cynomorium coccineum* L. (2n = 28). *Cynomorium* spp. reproduce by seeds that look like spores of the microbes and produced in huge amounts that fall on the soil and disseminate far distance by wind currents away from production sites. *Cynomorium coccineum*, the only member of the Saxifragales family Cynomoriaceae, is a root holoparasite that occurs in water-stressed habitats from western China to the Canary Islands. It parasitizes at least 10 angiosperm families from different orders, some of them only in parts of its range. This parasitic species is usually found on the roots of Nitrariaceae, Tamaricaceae, and Chenopodiaceous shrubs located in dry, rocky, or sandy soils of extreme desert environments (Cui *et al.*, 2013). *Nitraria tangutorum* and *Nitraria sibirica* are the preferred hosts (Wan and Chen, 2000). It emerges once a year to form an inflorescence up to 40 cm tall with thousands of highly reduced flowers. Its range extends from the Canary Islands (Lanzarote) and Mauritania through the Eastern Mediterranean in Saudi Arabia, Dubai and the Iranian region to the Mongolian deserts in western China (Bellot *et al.*, 2016). It is found in Tunisia, Bahrain, Jordan, Spain, Portugal, Italy, and Malta. It grows in sandy and rocky soils, usually in the desert or sub-desert habitats (Ben Attia *et al.*, 2019).

The hosts of *Cynomorium* are known as trees or shrubs among which are *Coffea*, *Salsola vermiculata*, *Atriplex* spp. including *Atriplex dimorphostegia* Kar. and Kir., *Atriplex halimus* L., *Atriplex rosea* L. and *Atriplex tatarica* L. (Gill and Hawksworth, 1961; Musselman; 1980; Riches and Parker, 1995; Hawksworth and Wiens, 1996), *Salvia* spp., *Ferula communis*, *Haloxylon persicum* (Qasem, 2011), *Tamarix* spp., *Nitraria retusa* and on *Retama raetam* trees in different Arab countries and others including Southern Europe. *Cynomorium coccineum* has been reported on *Tamarix tetragyna* (Pazy *et al.,* 1996) and is generally parasitizes species from Chenopodiaceae (usually *Atriplex* or *Salsola*), Plumbaginaceae such as *Limonium*, Tamaricaceae (*Tamarix*), Frankeniaceae (*Frankenia*), Cistaceae, Fabaceae, Asteraceae. Nitrariaceae including *Nitraria* spp. and Zygophyllaceae such as *Peganum harmala* in Asia and the Mediterranean. In the Mediterranean, *C. coccineum* was found on different halophytic plants, particularly *Atriplex halimus*, *Obione portulacoides*, and *Inula crithmoides* (Rinaldi and Rescigno, 2015). It has also been reported on Tamaricaceae (*Reaumuria*), Zygophyllaceae (*Zygophyllum*), and *Salsola* as in Central Asian (Teryokhin *et al.*, 1975; Chen and Funston, 2007; Yang *et al.*, 2012; Cui *et al.*, 2013; Zhang *et al.*, 2015). It was also found on plants including *Haloxylon salicornicum*, *Salsola vermiculata* (Qasem, 2011) and *Retama raetam*. *Cynomorium coccineum* occurs in saline areas, on compacted sands, coastal sands and dunes and along roadsides, from sea-level up to 200 m altitude. It is a holoparasite occurring on the roots of a wide variety of shrubs, such as *Atriplex, Cistus, Frankenia, Halimione, Inula, Limonium, Medicago, Melilotus, Myrtus, Parapholis, Pistacia, Salsola* and *Tamarix* in Africa (Ben Attia *et al.*, 2019). In Asia, it is common on *Nitraria sibirica* Pallas.

Cynomorium coccineum subsp. *coccineum* (sometimes referred to as *Cynomorium coccineum*) was found in Mediterranean regions, especially on sand dunes of North Africa. It is reported from Lanzarote in the Canary Islands and Mauritania through Tunisia and Bahrain in the south; Spain (including Andorra but not the Balaerics), Portugal, Southern Italy, Sardinia, Sicily, Gozo and Malta(www.Maltawildplants.com - (2002-2018). It has also been mentioned from Kuwait, Bahrain, Qatar, the United Arab Emirates, Oman and Saudi Arabia and in the Eastern Mediterranean (Royal Botanic Garden, 2010). Its range extends as far east as Afghanistan, Saudi Arabia and Iran. It attacks the roots of a few coastal-rupestral plant species, including *Atriplex lanfrancoi* but possibly switches onto roots of *Limbarda crithmoides* and *Crithmum maritimum* too.

Cynomorium songaricum was found attacking *Nitraria* sp. (Luo *et al.*, 1993). Grows at high altitudes in Central Asia and Mongolia. It is extensively collected as a herbal remedy for illnesses, including sexual worries and nocturnal emissions (Nickrent *et al.*, 2005; Shu, 2007). In southwest

Tunisia, it parasitizing *Haloxylon salicornicum, Salsola vermiculata*, and *Retama raetam* while in the Mediterranean Basin where the bioclimate is less arid and is characterized by rainy winters and warm/dry summer, *C. coccineum* is found to grow on different halophytic plants, particularly *Atriplex halimus, Obione portulacoides*, and *Inula crithmoides* (Rinaldi and Rescigno, 2015).

ECONOMIC IMPORTANCE

Cynomorium coccineum was called fugues melitensis (Maltese mushroom) during the Middle Ages (SIUC, 2001). The whole plant appears reddish-brown, dark upon drying; hence it was believed that it would be useful to treat ailments of the blood. On top of that, the phallic shape meant the plant could also be used to treat sexual problems and for this reason, *Cynomorium* was a highly valued commodity (Mifsud, undated; Qasem, 2020).

The parasite stem contains different chemicals, including anthocyanins glycosides, triterpenoid saponins, and lignans (Dharmananda, 2011; Qasem, 2020). *Cynomorium coccineum* subsp. *coccineum* from Sardinia was found to contain gallic acid and cyanidin-3-O-glucosideas the main constituents (Zucca *et al.*, 2013; Cusimano and Renner, 2019). The plant is used medicinally since important in cases of kidney deficiency characterized by impotence, cold pain in the loins and knees, female barrenness, and constipation. It is believed that it would be useful to treat ailments of the blood. A component in sexual enhancement herbal products, it works in men and women and sometimes provides results within hours, and continues over several days of use. It is effective against dysentery, bleeding during pregnancy, nose and bottom bleeding, is eaten as a liver remedy, and heals wounds and digestive system ulcers. The complete plant is an aphrodisiac, seminal generative and a general tonic. It is a cholagogue, and strengthens relaxed joints. The plant's extract is hypotensive and is used for impotence (its current main application in commercial products), and there are also findings that the herb extract inhibits HIV, lowers blood pressure, and improves blood flow in laboratory experiments. The plant is used in treating apoplexy, venereal disease, high blood pressure, vomiting, irregular menstrual periods, and as a contraceptive and toothpaste (Qasem, 2020).

In folk medicine, it is used as an astringent and aphrodisiac and increases the sperm count. It improves the percentage of live sperm and their motility, decreases the number of abnormal sperm and lowers blood pressure in dogs. It is a hair tonic and it has been mentioned that pulp powder in a mixture with honey is beneficial for impotence and deficiency in sperm liquid, while the plants extract increased immature rat castrates and the tube size of sperm production, and thus increased the number of sperms produced.

Cynomorium is available for human consumption in emergencies, famines, and in specific conditions. and it is currently used in herbal infusions and other food preparations (Cui *et al.*, 2013; Jin *et al.*, 2014; Qasem, 2020).Watson and Dallwitz (1992) indicated that C. *coccineum* is regarded as a famine food along with the host species it parasitizes. It has been mentioned as an emergency food cited in the Bible. The host plants- probably *Atriplex dimorphostegia, Atriplex halimus, Atriplex rosea,* and *Atriplex tatarica* were also eaten for their soft, mucilaginous leaves (http://www.hort.purdue.edu/new-crop/faminefoods/ff_families/CYNOMORIACEAE.html; Qasem, 2011; 2020), the stems are eaten raw or roasted like maize. The stem is cooked and mixed in some foods, giving it a red colour and an appetizing bitter flavour. The powdered stem is used as a spice on meat during its preparation. It is used as a red dye. The red coloured extract is used for dyeing clothes (Mahjoub, 2018). The plant is also used in leather tanning and is important in the industry; its dye is used for colouring jewelry and by the Bedouins for stamping their cattle (Qasem, 2020). The same people use the plant for anti-bile fluid retention. In the underground part of *Cynomorium*, there is a bitter sweet, which is edible. The juice of the swollen underground part of the stem base is used as a drink. However, the plant is of a strong bitter taste, and it is normally fed on during famines. It has been reported to harm the lung and cause skin roughness. A contradiction exists in the literature on its potential for grazing. While some reports indicate it could possibly be used fresh in animal feed, others reported the plant as being unpalatable.

CONTROL

Reports on control methods of *Cynomorium* spp. are lacking or very limited, which may be due to that the parasite is not found attacking species of economic importance but mainly forage or rangeland plants in the desert and some wild salty species in other places. Hand pulling of the parasite is widely practiced in different countries since is collected for different uses in medicine, food, and industry.

The larvae of the insect identified as *Eumerus* bulb flies (Syrphidae) were found feeding on *Cistanche* and then transfer to *Orobanche*, as well as to *Cynomorium*,

CYNOMORIACEAE (BALANOPHORACEAE) IN JORDAN (*CYNOMORIUM COCCINEUM*)

GENERAL BACKGROUND

Qasem (2011) reported only one species of Cynomoriaceae (*Cynomorium coccineum*) in Jordan that attacks 5 plant species. Hosts belong to the families Chenopodiaceae (3 species), Labiatae (1 species) and Umbelliferae (1 species) (Table 3.1.1)

Table 3.1.1. *Cynomorium cocceneum*, hosts and their families, status and intensity of infection in Jordan.

Host species	Family	Status	Intensity of infection	Prevalence
Atriplex leucoclada Boiss.	Chenopodiaceae	W	High	Limited
Ferula communis L. (NC)	Umbelliferae	W	Light	Limited
Haloxylon persicum Bunge ex Boiss. and Buhse	Chenopodiaceae	C	Light	Rare
Salsola vermiculata L.	Chenopodiaceae	W	High	Limited
Salvia palaestina Bentham	Labiatae	W	Light	Rare

W: wild, C: cultivated
Rare: only on few plants in 1-2 sites of a biogeographical region
Limited: on many plants localized in certain locations of 1 or 2 biogeographical regions
NC: Attachment was not confirmed
Source: modified after Qasem, 2011

Photos of *Cynomorium coccineum* at different growth stages and this parasite and its hosts in Jordan are illustrated in plates (1-8).

SOME ILLUSTRATIONS ON *CYNOMORIUM COCCINEUM* AND ITS HOSTS IN JORDAN

Plate 1. *Cynomorium coccineum* at flowering stage

Plate 2. *Cynomorium coccineum* showing the phallic shape

Plate 3. *Cynomorium coccineum* on *Ferula communis*

Plate 4. *Cynomorium coccineum* on *Atriplex* sp.

Plate 5. *Cynomorium coccineum* on *Haloxylon persicum*

Plate 6. *Cynomorium coccineum* on *Salsola vermiculata*

Plate 7. *Cynomorium coccineum* on *Salsola vermiculata*

Plate 8. *Cynomorium coccineum* on *Salvia* sp.

Plate 9. Wadi Al-Al in Jordan where *Cynomorium coccineum* is found in mountains

REFERENCES

Abd El-Rahman, H.A., El-Badry, A.A., Mahmoud, O.M., Harraz, F.A., The effect of the aqueous extract of *Cynomorium coccineum* on the epididymal sperm pattern of the rat. *Phytother. Res.*, 1999, 13, 3, 248–250.

Angiosperm Phylogeny Group. *An update of the Angiosperm Phylogeny Group Classification for the Orders and Families of Flowering Plants: APG IV, Bot. J. Linn. Soc.*, 2016, 161 (2), 105–20, doi: 10.1111/boj.12385.

Bellot, S., Cusimano, N., Luo, S., Sun, G., Zarre, S., Gröger, A., Temsch, E., Renner, S.S. Assembled plastid and mitochondrial genomes, as well as nuclear genes, place the parasite family cynomoriaceae in the saxifragales. *Genome Biol.Evol.,*2016, 8, 2214–2230.

Ben Attia, I., Zucca, P., Marincola, F. C., Piras, A., Rosa, A., Chaieb, M., Rescigno, A. The modern use of an ancient plant: exploring the antioxidant and nutraceutical potential of the Maltese mushroom (*Cynomorium coccineum* L.). *Antioxidants*, 2019. 8, 289; doi: 10.3390/antiox8080289.

Chen, G. L., An, T. Y., Jin, S. W., Chen, Q. Study on seed characters and vigor of *Cynomorium songaricum* Rupr. *Seed*, 2011, 30, 21–23. 10.16590/j.cnki.1001-4705.2011.01.053.

Chen, J., Funston, A. M. Cynomoriaceae. In: Wu, Z. Y., Raven, P. H., Hong, D. Y., (eds.), *Flora of China*, vol. 13, Clusiaceae through Araliaceae, Science Press, Beijing, China; Missouri Botanical Garden Press, St. Louis, MO, USA. 2007, pp. 434.

Christenhusz, M. J. M., Byng, J. W. The number of known plants species in the world and its annual increase. *Phytotaxa.*, 2016, 261 (3), 201–217. Doi: 10.11646/phytotaxa.261.3.1.

Cronquist, A. *An Integrated System of Classification of Flowering Plants*. Bronx, New York, Columbia Univ. Press; 1262. 1981.

Cui, Jin-Long, Vijayakumar, V., Zhang, G. Partitioning of fungal endophyte assemblages in root-parasitic plant *Cynomorium songaricum* and its host *Nitraria tangutorum*. *Front. Microbiol.*, 2018, 9, 666. DOI: 10.3389/fmicb.2018.00666.

Cui, Z., Guo, Z., Miao, J., Wang, Z., Li, Q., Chai, X., Li M., The genus *Cynomorium* in China: An ethnopharmacological and phytochemical review. *J. Ethnopharmacol.*, 2013, 147, 1, 1–15.

Cusimano, N., Renner, S.S. Sequential horizontal gene transfers from deterrent hosts in a widespread Eurasian parasitic plant, *Cynomorium coccineum. Am. J. Bot.,* 2019,106, 679–689.

Dharmananda, S. *Cynomorium* – Parasitic plant widely used in traditional medicine, 2011. Institute for Traditional Medicine, Portland, Oregon, USA. Available at: http://www.itmonline.org/arts/cynomorium.htm.

Dong, S. L., Chen, N. L., Hen, N. L., Zhang, Y., Li, C. X., Gao, H. N. Preliminary study on dormancy and germination of *Cynomorium songaricum* seed. *Arid Zone Res.*, 2010, 28 655–659. 10.13866/j.azr.2011.04.010.

Duke, J.A., Duke, P.A.K., du Cellier, J.L. *Duke's Handbook of Medicinal Plants of the Bible*, CRC Press Taylor and Francis Group: Boca Raton, FL, USA, 2008, 552.

Gill, L. S., Hawksworth, F. G. *The Mistletoes, A Literature Review*. U.S.D.A. *Technical Bulletin* 1961,1, 242 pp.

Harborne, J.B., Saito, N., Detoni, C.H. Anthocyanins of *Cephaelis, Cynomorium, Euterpe, Lavatera* and *Pinanga. Biochem. Syst. Ecol.*, 1994, 22, 835–836.

Hawksworth, F. G., Wiens, D. *Dwarf Mistletoes: Biology, Pathology and Systematics*. USDA Forest Service, Washington D.C. 1996.

Hooker, J.D. On the structure and affinities of Balanophoraceae. *Trans Linn Soc. London*, 1856, 22,1-68.

Jin, S.W., Chen, G.L., Du, J.J. M., Wang, L.H., An, T.Y., Antioxidant properties and principal phenolic composition of *Cynomorium songaricum* Rupr. *Int. J. Food Prop.*, 2014, 17, 13–25.

Kruh, L. L., Lahav, T., Abu-Nassar, J., Achdari, G., Salami, R., Freilich, S., *et al.* Host-parasite-bacteria triangle: the microbiome of the parasitic weed *Phelipanche aegyptiaca* and tomato-*Solanum lycopersicum* (Mill.) as a host. *Front. Plant Sci.,* 2017, 8, 269. 10.3389/fpls.2017.00269.

Kuijt, J. *The Biology of Parasitic Flowering Plants*. Berkeley, CA, University of California Press, 1969, 246.

Luo, G. H., Wang, J., Yan, X., Zhang, Y., Zhang, G. X., Wang, J. Q. Differences of *Cynomorium songaricum* seed quality and mutual parasitism in different host plants. *Chin. J. Chin. Mater. Med.,* 2013, 38, 3432–3437. 10.4268/cjcmm20132005.

Luo, M., Su, G.E., Liu, J.H., Ao, T.G. Study on the seed germination of *Cynomorium songaricum* without host. *Grassland of China,* 1993, 2, 62-64.

Mahjoub, J. Extraction of eco-friendly natural dyes from *Tradescantia pallid, Purpurea* and *Cynomorium coccineum* growing naturally in Tunisia. *Trends Textile Eng Fashion Technol.,* 2018, 1(1). TTEFT.000502.

Mifsud, S. (undated). Plants with orange to brown flowers. Available at: www. Malta Wild Plants. com - (2002-2018)

Musselman, L. J. The biology of *Striga, Orobanche*, and other root-parasitic weeds. *Annu. Rev. Phytopathol.*, 1980, 18, 463-489.

Nickrent, D.L., Der, J.P., Anderson, F.E. Discovery of the photosynthetic relatives of the "Maltese mushroom" *Cynomorium. BMC Evol. Biol.,* 2005,5, 38 doi:10.1186/1471-2148-5-38.

Pazy, B., Plitmann, U., Cohen, O. Bimodal karyotype in *Cynomorium coccinum* L. and its systematic implications. *Bot. J. Linn. Soc.*, 1996, 120, 279-281.

Qasem, J. R. *The Coloured Atlas of Medicinal and Aromatic Plants of Jordan and Their Uses.* Volume Two. Cambridge Scholars Publishing. Newcastle, UK. 2020, 489PP. ISBN 1527549216, 9781527549210.

Qasem, J.R. Parasitic flowering plants of woody species in Jordan. *Eur. J. Plant Pathol.,* 2011, 131, 143–155, DOI 10.1007/s10658-01-9794-2.

Riches, C. R., Parker, C. Parasitic plants as weeds. In: Press, M. C., Graves, J. D., (eds.), *Parasitic Plants*, London, Chapman and Hall, 1995 pp. 226-255.

Rinaldi, A.C., Rescigno, A. The wondrous *Cynomorium*. HAUSTORIUM Parasit. *Plants Newsl.*, 2015, 67, 11–12.

Royal Botanic Gardens. *World Checklist of Selected Plant Families*. Kew. 2010.

Shu, S. Y. *Flora of China*, 2007, 13, 434. Available at: Http://flora.huh.harvard.edu/china/PDF/PDF13/Cynomorium.pdf. (Accessed 14 Feb. 2021).

SIUC/College of Science /Parasitic Plant Connection/CynomoriaceaeURL: http://www.science.siu.edu/parasitic-plants/Cynomoriaceae/index.html. Last updated: 07-July-2001 / dln.

Takhtajan, A. *Diversity and Classification of Flowering Plants.* New York, NY, Columbia University Press; 1997, 643.

Terekhin, E. S. Development of microsporangium, pollen grains, ovule, embryo sac in *Cymonorium* (Cymomoriaceae). *Bot. Zhurn.* (*Leningrad*), 1975, 60, 153-162.

Teryokhin, E.S., Nikiticheva, Z.I., Yakovlev, M.S. Development of the seed, endosperm and embryo in *Cynomorium songaricum* Rupr. (Cynomoriaceae). *Botanicheskii Zhurnal*, 1975, 60 (11), 153–162.

UBC Botanical Garden: *Botany Photo of the Day Archived 2011-06-22 at the Wayback Machine*, 26 Feb 2008 (Accessed 19 April 2011)

Wan, W. H., Chen, J. R. *Cynomoriaceae.* In*: Flora Reipublicae Popularis Sinicae*. Beijing: Science Press. 2000.

Watson, L., Dallwitz, M. J. *The Families of Flowering Plants: Descriptions, Illustrations, Identification, and Information Retrieval.* 1992. Version: 1^{4th} December 2000. http://biodiversity.uno.edu/delta/'.

Yang, Y., Yi, X., Peng, M., Zhou, Y. Stable carbon and nitrogen isotope signatures of root-holoparasitic *Cynomorium songaricum* and its hosts at the Tibetan plateau and the surrounding Gobi Desert in China. *Isotopes Environ. Health Stud.,* 2012,48, 483–493.

Zhang, M. L., Temirbayeva, K., Sanderson, S. C., Chen, X. Young dispersal of xerophil *Nitraria lineages* in intercontinental disjunctions of the old world. *Sci. Rep.,* 2015,5, 13840.

Zucca, P., Rosa, A., Tuberoso, C.I.G., Piras, A., Rinaldi, A.C., Sanjust, E., Dessì, M.A., Rescigno, A. Evaluation of antioxidant potential of Maltese mushroom (*Cynomorium coccineum*) by means of multiple chemical and biological assays. *Nutrients*, 2013, 5 (1), 149–61. Doi: 10.3390/nu5010149. PMC 3571642. PMID 23344249.

Websites

http://www.hort.purdue.edu/new-crop/faminefoods/ff_families/CYNOMORIACEAE.html
www.Maltawildplants.com - (2002-2018)

SUMMARY AND CONCLUSION

Parasitic flowering plants are of a widespread significance in different parts of the world. Some important genera are common in the Middle East countries and exert a major threat to many economic crops and forestry species. These include a large number of root and stem parasitic plants, among which *Orobanche*, *Cuscuta*, *Viscum* and *Loranthus* are widely spread and represent a major threat to agriculture. Species of these genera exhibit a wide variation in their growth habit and habitats; they are also different in the size of their host range, many of which can attack both cultivated and wild species. Reviewing their ecological, biological and physiological requirements and behaviour revealed great differences between species in their responses. Ecological and biological factors affecting the germination, growth, and development of these parasites were reviewed. Possible control measures for each species were carefully considered and analysed. This review indicated some promising herbicides, synthetic and natural chemicals, and emphasized the importance of some physical, mechanical, agricultural and biological control measures. However, an integrated control approach has been strongly recommended by different workers for parasitic weed control.

In Jordan, among parasitic genera recorded in the country, root parasitic species presented in this book volume are important and of main concern since they represent a threat to both agro- and ecosystems. Broomrapes (*Orobanche* spp.) are widely spread and serious challenge and danger to agriculture, with 11 *Orobanche* species are documented in the country, but seven species cause serious damage to vegetables, field crops, ornamentals, fruit trees, and forestry species. The total number of host species these attack was 86 that belong to 24 plant families. *Orobanche ramosa, O. aegyptiaca, O. cernua, O. crenata, O. palaestina, O. minor* and *O. schultzii* are of main importance since parasitizing a wide range of vegetables, fruit trees and other economic crops. In addition, *Osyris alba* is second in its economic damage to fruit trees and some forestry species, with 18 host species recorded and 5 others may possibly be attacked. However, this parasitic species attacks 14 fruit tree species, including olive, grapes, almonds, and plums. Other root parasitic genera in the country, including *Cistanche* spp., *Cynomorium coccinum*, and *Thesium humile* are less economical in agricultural land but mostly attack forage and wild shrubs in dry and saline habitats, while *Thesium humile* was found attacking onion and barley but of extremely less prevalence and importance.

GLOSSARY

Glossary of Scientific Terms

- **Allelochemicals**: Natural chemicals are usually liberated as volatiles, exudates, leachates from plants, or their decayed residues in nature and can inhibit or stimulate germination, growth, and development of other plant species or the general biosystem. Some of the chemicals released may have an autopathy effect on the same plant that produces them. Allelochemicals are usually bi-products or secondary metabolites that have no exact function in the plants.
- **Allelopathy**: Any direct or indirect harmful or stimulatory effect of one plant (including microorganisms) on another through the production of chemical compounds that escape into the environment.
- **Angiosperm**: Flowering plants of a large group that comprises those that have flowers and produce seeds enclosed within a carpel. This group includes herbs, shrubs, grasses, and most trees. Angiosperms produce seeds encased in edible fruits, but also plants might thought as fruits, such as maple seeds, acorns, beans, wheat, rice, and corn.
- **Annual plant**: Plants complete their life cycle and live for less than a year from germination until seed production and death. They are herbs reproduce by seeds and can be included into two groups. Winter annuals are plants that germinate late in autumn or during the winter grow vegetatively during the winter and spring time and by the end of the spring, or early in the summer, they flower and produce their seeds the n die. Their seeds remained dormant in the soil throughout the summer and autumn and then germinate in the next season. The second group is summer plants at which these start germination late in the spring or early in the summer grow vegetatively during the summer and most of the autumn; then, at the end of autumn, they flower, produce seeds, and then die. Their seeds remain dormant in the soil until the next season.
- **Appressorium**: A structure appeared in the mistletoe germinated seed on host branches; it is developed from below the holdfast penetrating the host bark through mechanical pressure and analytic enzymes.
- **Autoparasitism**: A kind of parasitism found in *Viscum cruciatum* in which seeds of this species dropped on their own shoots, germinate, parasite seedlings grow, and start sucking directly from the main parasite itself.
- **Biennials**: Plants live only for two years. In the first year, they grow vegetatively in a rosette growth form and basal leaves, while in the second year; they produce flowers and a huge number of seeds. However, if the condition for flowering were not fulfilled, then the plant can live for several years but remain dormant at vegetative growth.
- **Bimodal karyotype**: Refers to a symmetrical karyotype that is composed of two distinct classes of chromosomes as determined from their size. Examples of such karyotypes are found in certain genera of Liliales, such as *Aloe*, *Gasteria*, *Yucca*, *Agave*, etc. Bimodal karyotypes are characterized by the presence of two sets of chromosomes of contrasting size
- **Biocontrol agent**: Living bioagents used could be an insect, fungi, bacteria, nematodes, animals, and even allelopathic plants. They are used as bioagents to serve as natural enimies for different pests, including parasitic weeds, and through biological control programs or in integrated parasitic weed control programs.
- **Biological control**: The use of natural enemies for pest control. These are specific and not harmful to crop plants. They generally include insects and fungi used for parasitic weed control.

- **Botanical pesticides**: Natural chemicals mainly of plant origin. They are produced by plants and may be prepared in different forms and used against different agricultural pests through allelopathy phytotoxic
- **Catch species**: These are host plants (possible fodder harvest or green manure) that induce parasite seed germination and get parasitized. They are sometimes grown to exhausted parasite seed banks since they are destroyed after parasite attachment on their roots but before they emerge and produce seeds. They are regarded as crops to sacrify for parasitic weed management.
- **Chemigation**: Application of pesticides, especially herbicides, through irrigation. Different herbicides are added through this method of application and some proved effective in controlling certain parasitic weeds such as *Orobanche*.
- **Chemotropic response orchemotropism**:The growth of organisms navigated by chemical stimulus from outside of the organism. It has been observed in bacteria, plants and fungi. A chemical gradient can influence the growth of the organism in a positive or negative way. Positive growth is characterized by growing towards a stimulus, and negative growth is growing away from the stimulus. Chemotropism is slightly different from chemotaxis, the major difference being that chemotropism is related to growth, while chemotaxis is related to locomotion.
- **Conditioning period**: A period during which *Orobanche* seeds became ready to respond for stimulants released from the host root system, and parasite seeds became more responsive to germination stimulants. Conditioning**,** in physiology, is a behavioral process whereby a response becomes more frequent or more predictable in a given environment as a result of reinforcement, with reinforcement typically being a stimulus or reward for the desired response. A process of behavior modification by which a subject comes to associate the desired behavior with a previously unrelated stimulus.
- **Crop rotation with trap/catch crop**: A technique through which host crop is grown in rotation after trap or catch species that exhaust parasite seed bank in the soil.
- **Dioecious hemiparasite**: A parasitic plant at which has male and female plants separated in two plants. An example is *Osyris alba* that lives as facultative, has chlorophyl, do photosynthesis and produces some carbohydrates and can grow separately or parasitizing other suitable plant species.
- **Dioecious plant**: A plant species has both male and female plants, but one has "male" **flowers** and the other "female" flowers. Some well-known dioecious plants include holly, asparagus, dates, mulberry, ginkgo, persimmons, currant bushes, juniper bushes. It is opposite to the monoecious plants that produce male and female flowers on a single plant.
- **Dormancy**: Failure of seeds to germinate when conditions are suitable for germination. This is called genetic or innate dormancy, while the other two types are also known. These are secondary or induced dormancy and enforced dormancy.
- **Drupe fruit**: Also called stone fruit. It is an indehiscent fruit in which an outer fleshy part (exocarp, or skin; and mesocarp, or flesh) surrounds a single shell (the pit, stone, or pyrene) of hardened endocarp with a seed (kernel) inside.
- **Epicortical roots**: In some Loranthaceae, roots develop from the cortex of the stem to form secondary points of attachment to the host**.** Epicortical roots bearing secondary haustoria are conspicuous features of many mistletoe genera. Mistletoes produce epicortical runners, roots that grow in vine-like fashion outside the host branch, sending down relatively simple secondary haustoria at regular intervals (below). Epicortical runners are more common in mistletoes of humid forests. However, the direction of

evolution may have been towards a more efficient primary haustorium and consequent reduction and loss of the epicortical runners.

- **Epigeal stage**: The stage during which the vegetative organ grows rapidly and produce reproductive organs above the soil, indicating that the germination of a plant takes place above the ground. An example of a plant with epigeal germination is the common bean (*Phaseolus vulgaris*). The opposite of epigeal is hypogeal (underground germination). Epigeal is also not the same as hypogeal germination; both epigeal and hypogeal plants will grow differently. However, seedling growth can be divided into two types: epigeal and hypogeal. Angiosperm seedlings fall into two main classes with respect to the fates of their cotyledons during the growth of the axis. Seedlings that raise their cotyledons above the soil surface are said to be epigeal, while those whose cotyledons remain in the soil are termed hypogeal. Onion (*Allium*) is the hypocotyl, located below the cotyledonary node In most epigeal seedlings, both the hypocotyl and epicotyl elongate. In contrast, only the epicotyl elongates in hypogeal seedlings, such as in pea (*Pisum*). The seeds of members of the grass family are also hypogeal,
- **Epiparasitism**: This is concerned with certain species such a*s Viscum capitellatum* that appeared restricted to specific Loranthaceae species as only hosts for this parasite. Epiparasitism is known for mistletoes of both Loranthaceae, Viscaceae and also Santalaceae represented by *Phacellaria* the obligate parasite of Loranthaceae of Southeast Asia. Epiparasites (also called hyperparasites) are parasitic plants that parasitize other parasitic plants of a different species.
- **Eradication**: Complete elimination of the targeted weed species and its propagules from the environment.
- **Facultative parasite**: a parasitic species that doesn't require a host to complete its life cycle such as *Osyris* and *Thesium* species. These parasites are able to do photosynthesis and produce carbohydrates throughout the entire life cycle and can grow separately, but when they find a suitable closer growing host, they attack it.
- **Facultative (hyperparasites)**: Facultative hyperparasitism involves the development of the progeny as either primary or secondary parasitoids. In fact, facultative hyperparasitism is a very aggressive behaviour of females that sting and kill the developing primary parasitoid (L5 larvae or pre-pupae or pupae) before ovipositing on it. A hyperparasite is a parasite whose host, often an insect, is also a parasite, often specifically a parasitoid.
- **Gene silencing**: An epigenetic modification of gene expression leading to inactivation of previously active individual genes or larger chromosome regions. Mechanisms responsible for repression of genes involve changes in levels of DNA methylation and alterations in covalent modifications of histone proteins, which lead to chromatin compaction, making genes inaccessible to the transcription machinery. It is a new approach recently developed and introduces a double-stranded (dsRNA) as an effective tool for suppressing gene expression through a process known as an interference (RNAi) in animals and post-transcriptional gene silencing (ptgs) in plants environmentally safe.
- **Geophytes:** A perennial plant that propagates through underground buds or has an underground storage organ such as a bulb, corm, or rhizome.
- **Germination stimulants**: These are mainly natural chemicals produced by host plant species and some of the nonhosts that can stimulate seed germination of parasitic weed such as *Orobanche* otherwise in the absence of these chemicals parasite germination won't occur. Trap species usually produce these chemical stimulants and stimulate

parasite seed germination but without being themselves attacked. Some synthetic chemicals can also induce seed germination of certain parasitic weeds.

- **Germination tube**: It is a primary growth, projected from the parasite germinated seed in response to host germination stimulants. It is mainly found for holo-root parasites such as Orobanche. It is sometimes called a radicle but is not a root, although it looks so. It grows toward the host root system attaching the parasite seedling to the host root and starts forming the haustorium on its tip, which penetrates the host root tissues and starts sucking water and nutrients and then the tube disintegrated and disappeared.
- **Haustorium (pl. Haustoria)**: An absorptive organ like a sucker produced by parasitic weed and inserted inside host tissues and form a bridge of conductive channel for nutrients, water and food materials from the host to its parasite.
- **Hemi- parasites (incomplete parasites)**: Description of a parasitic weed with chlorophyllpigments and can do photosynthesis, but for certain reasons, it can also attack a suitable neighboring plant species and parasitize it. These weeds may be found growing separately throughout their life cycle or parasitizing other plant species. Examples are *Osyris alba*and *Thesium humile*.
- **Herbicide selectivity**: It means that the herbicide affects a certain type or group of plants but not others. Selectivity is not absolute but rather depends on different factors, including plants, herbicides and environmental factors. Therefore, when the herbicide stunts, inhibitsor kills certain groups of weeds such as only broadleaved and not narrow-leaved weds or vice versa or it kills weeds but doesn't harm the crop, it is said to be selective herbicide.
- **Hermaphrodite**: An animal or plant having both male and female reproductive organs. Hermaphrodite vs. Intersex The two terms Hermaphrodite and Intersex have been used synonymously. Hermaphroditism**,** the condition of having both male and female reproductive organs. Hermaphroditic plants most flowering plants, or angiosperms, are called monoecious, or bisexual.
- **Holo-parasites (or complete parasites)**: Parasites that are devoid of chlorophyll (or nearly so); they are completely dependent on their hosts for organic carbon, nutrients, and water, and most are root parasites.
- **Host preference**: Refers to the choice of the most desirable host for optimal growth.
- **Host range**: Refers to the total number of different species that can be parasitized.
- **Hydrolytic enzymes**: Enzymes that catalyze the hydrolysis of a chemical bond. These tend to be useful in determining the lipid structure.
- **Hyperparasitism**: Mainly describes the association between different parasitic species (*cuscuta* and *cassytha*) that parasitize a variety of plants include woody root hemiparasites (eg. *Ximenia*, *santalum,*and others), and also as hyperparasites of mistletoes such as *Phoradendron* and *Struthanthus*. Hyperparasitism is the presence of one parasitic angiosperm upon another. It takes place when a parasite lives at the expense of another parasite, either on or inside it.
- **Hypersensitive reaction**: Hypersensitivity (also called hypersensitivity reaction or intolerance) refers to undesirable reactions produced by the normal immune system, including allergies and autoimmunity. Hypersensitivity reactions occur when the normally protective immune system responds abnormally, potentially harming the body. Various autoimmune disorders, as well as allergies, fall under the umbrella of hypersensitivity reactions, the difference being that allergies are immune reactions to exogenous substances (antigens or allergens), whereas autoimmune diseases arise from

an abnormal immune response to endogenous substances (auto antigens). A symptomatic reaction only occurs in sensitized individuals.

- **Hypogeal stage**: Plant growth stage that occurs below the soil level and lasting for 60 days in *Orobanche* during which the vegetative subterranean organ develops and food materials absorbed.
- **Inhibitory substances**: Any substance inhibits parasite seed germination, growth or development. It may cause abnormal growth of the parasite and failure to complete its life cycle normally. It can also be defined as a substance that slows or prevents a chemical or organic reaction.
- **Integrated control or Integrated Pest Management (IPM)**: It is an effective and environmentally sensitive approach to pest management that relies on a combination of common-sense practices. IPM takes advantage of all appropriate pest management options, including, but not limited to, the judicious use of pesticides.
- **Intercropping**: A multiple cropping practices involving growing two or more crops in proximity. It aims mainly to produce a greater yield and more crops in a given area by making use of resources or ecological processes that would otherwise not be utilized by a single crop. Intercropping is growing two or more crops next to each other at the same time.
- **Minimum tillage**: Also called conservation tillage, it is a soil conservation system like Strip-till with the goal of minimum soil manipulation necessary for successful crop production. It is a tillage method that does not turn the soil over. Soil cultivation is kept to the minimum necessary for crop establishment and growth, thereby reducing labor and fuel costs and damage to soil structure.
- **Mini-rhizotron**: A system made up of durable, transparent tubes installed in the ground and a light-weight cylindrical imaging device that is lowered into the tubes to collect high-resolution images and track root growth season to season. Minirhizotrons allow accurate and non-destructive sampling of fine root production. Minirhizotron cameras are systems developed to obtain images from the soil biological activity (roots growth in particular
- **Monoecious plant**: A plant carries both male and female flowers or has flowers on every plant containing both male and female reproductive components. It is in contrary to a dioecious plant that has either male or female flowers, but not both.
- **Monophagous insect**: An insect restricts itself to a single host species. it is a consummate specialist, adapting its behavior and physiology to a single nutritional resource. It is feeding upon only one species or genus of plants
- **Mycoherbicides**: These are herbicides based on fungi. Mycoherbicides are developed from plant pathogenic fungi that occur naturally in the environment. As a bioagent, mycoherbe works by producing toxic compounds that dissolve the cell walls of targeted plants. These herbicides can reproduce themselves and linger in the soil for many years to destroy replanted crops. Mycoherbicides are mainly used to prevent and control the spread of such worst parasitic weeds as *Orobanche, Phelipanche, Striga*, and *Cuscuta*.
- **Neotropics**: A zoogeographical or phytogeographical region comprising Central and South America, including the tropical southern part of Mexico and the Caribbean.
- **Obligate parasites**: Parasites are completely dependent on the host in order to complete their life cycle. Over time, they have evolved so that they can no longer exist without the existence of the host. This means that they usually will not cause serious harm to the host,

as the parasites need the host to survive so that they can survive unless the host's death is needed for the transmission of the parasite. Obligate parasitism can be found in many different types of organisms, like plants, animals, fungi, bacteria, and viruses. Head lice are obligate parasites; if removed from the human scalp, they will soon die. Epiparasitism is a parasite that parasitizes another organism that is also a parasite. Epiparasites are also called hyperparasites or secondary parasites such as *Viscum* and *Loranthus* spp.

- **Obligate holoparasites**: These are completely host-dependent to flower and seeding or fruiting such as *Orobanche* and *Cuscuta*. They don't exist in the absence of host plants since devoid of chlorophyll.
- **Oligophagous**: Insects feeding on a limited number of foods, usually within one taxonomic family. These are feeding upon a limited variety or number of foods, as certain caterpillars. Feeding on a limited number of foods, usually within one taxonomic family, such as an oligophagous pest of leguminous plants.
- **Paleotropics or the Paleotropical Kingdom (Paleotropis)**: It is a floristic kingdom comprising tropical areas of Africa, Asia, and Oceania (excluding Australia and New Zealand). The Paleotropical, or Afro-Tethyan, is divided into Afrotropical, which includes continental Africa south of the Sahara and southwestern Arabia, and the Oriental, which includes tropical southern and southeastern Asia.
- **Parasitic weeds:** These are weeds that depend totally or partially upon their hosts for food and/or water. These may not be able to obtain growth factors by themselves from the environment (holo-parasites) but only through other plants (hosts) or what they absorb by themselves is not enough to accommodate them for the entire life cycle (hemiparasites). Parasitic weeds could be root-parasites or shoot-parasites.
- **Perennials**: Plants that complete their life cycles in more than two years and could be extended for more than 10 years. These either reproduce by seeds like annual plants or creeping reproduce mainly by vegetative organs. Perennial weeds are difficult to control.
- **Phanerogamic plants**: These also parasitize the crop plants. Phanerogams are seed-bearing most advanced plants. They are also known as spermatophytes.
- **Phototropism**: One of the many plant tropisms or movements which respond to external stimuli. Growth towards a light source is called positive phototropism, while growth away from light is called negative phototropism. Phototropism is the ability of a plant, or other photosynthesizing organisms, to grow directionally in response to a light source.
- **Phytoallyxins**: Antimicrobial and often antioxidative substances produced by plants to combat infection by a pathogen (such as a fungus). They accumulate rapidly at areas of pathogen infection and include broad-spectrum inhibitors, and are chemically diverse with different types characteristic of particular plant species.
- **Phytohormons**: Naturally occurring small organic molecules or substances which influence physiological processes in plants at very low concentrations. They are chemical messengers that coordinate the cellular activities of plants.
- **Phytophagous insects**: Insects feed on green plants and use plants in several to many plant families. They include species that attack roots, stems, leaves, flowers, and fruits, either as larvae or as adults or in both stages. Leaf feeders may be external (exophytic) or they may mine the tissues, sometimes even specializing on a particular cell type. Phytophagous is often synonymous with "herbivorous", although the latter term is sometimes restricted to those species feeding on herbs. Commonly, species that use only one plant genus or species are called monophagous and species that use plants within a

tribe or family are called oligophagous. The term stenophagous is less commonly used and includes both of these.

- **Plant residues and soil cover:** A technique for weed control at which dead plant residues are either soil-incorporated or spread on the soil surface (surface mulch). Residues act physically by preventing weed seed emergence if the residue layer is thick enough and doesn't allow seedlings to emerge. They act chemically through the release of allelochemicals that inhibit weed seed germination or seedlings' growth.
- **Polyphygeous:** Feeding on many different kinds of food.
- **Preconditioning phase**: The phase during which the seed embryo is conditioning under suitable temperature and moisture. This period was considered important for leaching germination inhibitors from broomrape seeds while others suggested that germination stimulants or related substances accumulate during this period.
- **Prehaustoria**: A stage in parasitic weed growth between stimulant reception and radical attachment to the host plant surface and actual haustorium formation. In the first physical contact initiates an attachment phase, in which the parasitic epidermal and parenchyma cells begin to differentiate into a secondary meristem and develop the prehaustoria, it is also known as an adhesive disk. The penetration phase follows the attachment phase as prehaustoria develop into parasitic haustoria that penetrate the host stem through a fissure.
- **Prevention methods**: All measures prevent parasite introduction into a new parasite-free field or region. These include different precautions, sanitation and measures to avoid contamination of crop seeds or seedlings with parasite seeds.
- **Resistant species**: Plant species not affected by exposure to prevailing environmental stress conditions such as disease resistance, drought resistance or salinity resistance. Resistance is the inherited ability of a plant or a biotype to survive and reproduce following exposure to certain stress conditions,: herbicides. Herbicide resistance is mainly concerned with many weed species or biotypes that are not affected or harmed and continue growing and reproducing after being exposed to a dose of herbicide that is normally lethal to wild type.
- **Root parasites**: These are parasitic weeds connect to the root system of host plants through absorptive organs (haustoria) whether they are holo- or hemiparasites. Examples are *Orobanche, Osyris, Thesium, Cistanche* and *Cynomorium.*
- **Shoot or stem parasite**: Parasitic weeds subset themselves on stems or shoots of host plants. They can be holo- or hemiparasites. Examples are *Cuscuta, Viscum, Loranthus* and other.
- **Soil mulch**: Soil cover with a layer of material (plastic, paper, carton, organic manure/matter, living or dead mulch, stones...etc.) applied to the surface of the soil with aim include conservation of soil moisture, improving fertility and health of the plant. Mulch is used to retain moisture in the soil, suppress weeds, keep the soil cool, prevent frost heaving in winter, and make the garden bed look.
- **Soil solarization**: An eco-friendly method of using the sun's energy to control pest such as bacteria, insects, nematodes and weeds in the soil. The process involves covering the ground with a tarp, usually a transparent polyethylene cover, to trap the solar energy.
- **Stimulatory substances**: Any chemical or compound, whether synthetic or naturally produced, enhances seeds of parasitic weeds to germinate and break dormancy. Stimulants are mainly produced by host plants and can differ quantitively and qualitatively between host species and differences in their chemical nature. However,

some analogue chemicals have also been synthesized and produce as stimulating parasite seed germination and can be added as a pre-planting treatment to cause suicidal effects on parasite seedlings and exhaust parasite seed banks in the absence of host species.

- **Susceptible species**: A plant species that is highly sensitive to external treatment or stress factor caused by agricultural pest or pesticides mainly herbicides. It is injured or rapidly inhibited toxicity or injury symptoms or sometimes death after exposure to unsuitable conditions.
- **Symbiotic relationship**:Symbiosis is a close relationship between two species in which at least one species benefits. For the other species, the relationship may be positive, negative, or neutral. There are three basic types of symbiosis: mutualism, commensalism, and parasitism.
- **Systemic herbicides:** The chemical either soil or foliage applied on plants mainly to control weeds, absorbed through plant surface (shoots or roots) and translocates inside the plant from the site of application to the site of action where it shows activity against the weed. These herbicides may be general in action killing all vegetation, including crops, and are used where no vegetation is required or selective that kill broadleaved or narrow-leaved weeds but not both and do not harm crop plants.
- **Tolerant species**: Species that can live, tolerate (or survive within) a certain range of a particular factor but can't survive if there is too much or too little of the factor.
- **Trap species**: A false host that stimulates the parasite seed germination but does not support parasitism. Thus, it works in a suicidal way that leads to exhaustion of parasite seed banks when integrated with other methods of parasite management.
- **Tripartite epi- or hyperparasitic associations**, such as *scurrula ferrugineus* on *viscum articulatum* on *elytranthe barnesii* which was itself parasitic on *durio* . tripartite means has 3 parts. **Epiparasitic** plants specialized in arbuscular mycorrhizal fungi.
- **Tubercles**: A small knobby prominence or excrescence, especially on a plant or animal. The nodule is a small rounded projection or excrescence, as on a bone or on the surface of the body
- **Viscin**: A sticky material produced by the parasite *Viscum.*
- **Zero tillage**: No-till farming (also known as zero tillage or direct drilling) is an agricultural technique for growing crops or pasture without disturbing the soil through tillage. No-till farming decreases the amount of soil erosion tillage causes in certain soils, especially in sandy and dry soils on sloping terrain.
- **Zygomorphic flower**: A flower having only one plane of symmetry, as in a pea or snapdragon; bilaterally symmetrical having bilateral symmetry; divisible lengthwise into similar or symmetrical halves.

APPENDICES

List of Scientific Names

List of Illustrations

Made in the USA
Las Vegas, NV
15 November 2026

235ff5eb-6429-4aab-975a-82f639e3d6daR01